Guides to Professional English

Series Editor:
Adrian Wallwork
Pisa, Italy

For further volumes:
http://www.springer.com/series/13345

Adrian Wallwork

User Guides, Manuals, and Technical Writing

A Guide to Professional English

Springer

Adrian Wallwork
Pisa
Italy

ISBN 978-1-4939-0640-6   ISBN 978-1-4939-0641-3 (eBook)
DOI 10.1007/978-1-4939-0641-3
Springer New York Heidelberg Dordrecht London

Library of Congress Control Number: 2014939424

© Springer Science+Business Media New York 2014
This work is subject to copyright. All rights are reserved by the Publisher, whether the whole or part of the material is concerned, specifically the rights of translation, reprinting, reuse of illustrations, recitation, broadcasting, reproduction on microfilms or in any other physical way, and transmission or information storage and retrieval, electronic adaptation, computer software, or by similar or dissimilar methodology now known or hereafter developed. Exempted from this legal reservation are brief excerpts in connection with reviews or scholarly analysis or material supplied specifically for the purpose of being entered and executed on a computer system, for exclusive use by the purchaser of the work. Duplication of this publication or parts thereof is permitted only under the provisions of the Copyright Law of the Publisher's location, in its current version, and permission for use must always be obtained from Springer. Permissions for use may be obtained through RightsLink at the Copyright Clearance Center. Violations are liable to prosecution under the respective Copyright Law.
The use of general descriptive names, registered names, trademarks, service marks, etc. in this publication does not imply, even in the absence of a specific statement, that such names are exempt from the relevant protective laws and regulations and therefore free for general use.
While the advice and information in this book are believed to be true and accurate at the date of publication, neither the authors nor the editors nor the publisher can accept any legal responsibility for any errors or omissions that may be made. The publisher makes no warranty, express or implied, with respect to the material contained herein.

Springer is part of Springer Science+Business Media (www.springer.com)

Introduction for the reader

Who is this book for?

This book is intended for anyone whose job involves writing formal documentation (e.g. user manuals, technical reports, RFQs). It is aimed at non-native speakers of English, but should also be of use for native speakers who have no training in technical writing.

Technical writing is a skill that you can learn and this book outlines some simple ideas for writing clear documentation that will reflect well on your company, its image and its brand.

The focus of the book is thus on writing skills: it does not cover issues of layout and graphics.

Most of the rules / guidelines contained in this book are not exclusive to English, and can be applied to documentation in any language.

What are the typical difficulties of writing company documentation? What is the focus of this book?

The main issues are:

1. knowing who your readers are: their level of prior knowledge and the expectations they have while reading your document
2. using a consistent layout, format and style
3. writing clearly and concisely with no ambiguity

This book is part of series of books on using the English language at work. It thus focuses on the third point above. However, being able to write clearly and concisely with no ambiguity also implies incorporating the other two points as well.

What are the differences between technical writing and other kinds of writing?

This book covers technical writing. However, most of the rules and guidelines given are equally applicable to reports, emails, presentations and other kinds of formal company documentation. The aim of all these documents is to be as clear as possible and thus enable the reader to assimilate the information in the least time possible.

This kind of writing differs massively from the way you were probably taught to write at school, where long sentences and an abundance of synonyms were probably considered signs of good writing. But you will have no problem in quickly understanding the benefit of writing in short simple sentences that contain no redundant words and are unambiguous.

Remember you are writing to inform. You are not writing to impress a teacher or your boss.

All documents that you produce for your company are a reflection of the company's brand and image.

Mistakes indicate a lack of professionalism behind the document, and thus the company in general. For example, misspelling of a client's name or of your own products reflects badly on the whole company.

When users buy a product, most will know what the product is. What they really want to know is how to use to it, rather than read a description of what it is. This means that you need to:

- find out who your readers will be and how they going to use your document. This also means discovering if there will be different types of readers (managers, technical staff, end users) and adjusting the style accordingly

- ascertain what prior information and knowledge your readers have. Then decide what they need to know and give them only that. At the same time, avoid making assumptions. Unless you know the reader, you can't be sure that their knowledge of the subject is as advanced as yours

- get your ideas clear in your mind. If you're not sure exactly what it is you're writing about, then how can you expect your readers to understand? So, write a plan. Get everything into its logical order

- for all types of readers, the document must be easy to understand quickly and it should be an enjoyable process for them – they get what they want with minimal effort

- make your guide predictable. Readers should intuitively be able find what they want easily and quickly. The whole guide should be presented in the same way throughout. For example, procedures should always be described in the same pattern – a series of steps with the outcome clearly given at the end
- ensure that all documents of a similar nature look the same. Readers can thus learn how to use them. So the next time they receive a technical document from your company, they will know where to find everything

The result will be that reader's expectations are met. They won't get tired or frustrated reading your document. Your aim is to use up as little of your reader's time as possible. People don't read company documents (manuals, reports etc) for fun, so you need to make the act of reading your documents a satisfying experience.

Bear in mind that people consult a user manual as a last resort. They are unable to solve the problem by themselves, so they look at the manual for help. When they read the manual they may be stressed or angry, so they need to find a solution in the quickest time possible.

Finally, remember that a user manual is not a piece of sales literature. Below is an extract from the Introduction to a medical device for home use.

> Congratulations on your choice of XXX for safe, effective pain relief.... You will soon wonder how you ever managed without your XXX.

Such sentences rarely add value for users as they have already bought the product and so do not need further convincing. Limit such usage to an advertisement.

How is this book organized? How should I read the book?

The book is organized into three parts.

PART 1 STRUCTURE AND CONTENT

This gives details about the typical sections found in a user manual and the order they are usually found in. Through examples, you will learn best practices in writing the various sections of a manual and what content to include.

PART 2 CLEAR UNAMBIGUOUS ENGLISH

This gives guidelines on how to write short clear sentences and paragraphs whose meaning will be immediately clear to the reader with no ambiguity.

PART 3 LAYOUT AND ORDER OF INFORMATION

Here you will find guidelines to issues of style such as the use of headings, bullets, punctuation and capitalization.

PART 4 TYPICAL MISTAKES

This section is divided alphabetically covering grammatical and vocabulary issues that are typical of user manuals.

Before drafting your first manual, I suggest you read Parts 1 and 2 carefully. Then read the contents page of Parts 3 and 4, and check if there any points that you are not clear about.

All the examples given have been adapted from real manuals and brochures.

How are the examples presented?

Examples are shown in two ways in this book:

1. Indented and in a smaller narrower font.

2. In a table with YES and NO headings.

For examples of type 1), I will tell you whether the use of bold and italics is for the purposes of the book (i.e. to highlight particular usages or mistakes) or whether bold and italics were part of the formatting of the example in its original location.

For examples of type 2), all cases of bold are to highlight particular usages or mistakes.

Where do the examples in this book come from?

The examples in Part 1 have been adapted from a wide range of guides, instruction manuals and company literature for the following: a powered golf trolley, software for trading on electronic money markets, a steam iron, a tumble dryer, an electric oven, a washing machine, an intra-sonic pain relieving device, a digibox, personal banking literature, and a supercharger.

The examples in the rest of the book either come from software manuals or were invented for the purposes of this book. The invented sentences are examples in general English which are used to show how grammatical rules are applied and certain vocabulary terms are used.

We are not native English speakers. Is it worth getting a professional (native English-speaking) technical writer to look at our documents?

Yes. Even if you can't employ a technical writer on your staff, you can at least use their services whenever you draft a new kind of document. A technical writer will not only check the grammar, language, punctuation, and spelling of your documents but will also:

- suggest a style (formatting, pagination) for your company's docs
- follow international guidelines on writing documentation (e.g. the Chicago Manual of Style)
- rewrite in order to clearly communicate what the author has in mind
- restructure, where appropriate, the order in which information is given
- tailor information to the knowledge level of the audience
- apply a "one word – one meaning" approach
- ensure clarity for multi-national English readers / speakers
- ensure that product names are written correctly (names, spelling, capitalization)
- have discussions with authors to enhance understanding and clarity
- check any legal notices e.g. copyright info and trademark acknowledgements

Terminology and abbreviations used in this book

doc	document
info	information
you	the person reading this book and thus the writer of company documents
product	the product or service described in a user manual
user	the reader of a user manual
user manual	a technical document intended to help user's use a product or service. Also known as a 'user guide'

Other books in this series

There are currently five other books in this *Professional English* series.

CVs, Resumes, and LinkedIn
http://www.springer.com/978-1-4939-0646-8/

Email and Commercial Correspondence
http://www.springer.com/978-1-4939-0634-5/

Meetings, Negotiations, and Socializing
http://www.springer.com/978-1-4939-0631-4/

Presentations, Demos, and Training Sessions
http://www.springer.com/978-1-4939-0643-7/

Telephone and Helpdesk Skills
http://www.springer.com/978-1-4939-0637-6/

All the above books are intended for people working in industry rather than academia. The only exception is *CVs, Resumes, Cover Letters and LinkedIn*, which is aimed at both people in industry and academia.

There is also a parallel series of books covering similar skills for those in academia:

English for Presentations at International Conferences
http://www.springer.com/978-1-4419-6590-5/

English for Writing Research Papers
http://www.springer.com/978-1-4419-7921-6/

English for Academic Correspondence and Socializing
http://www.springer.com/978-1-4419-9400-4/

English for Research: Usage, Style, and Grammar
http://www.springer.com/978-1-4614-1592-3/

Contents

PART I Structure and Content of a Manual

1 TItle—table of contents—about—introduction—product overview—what's in the box 3
   1.1 Title 3
   1.2 Table of Contents 4
   1.3 About 5
   1.4 Introduction / Product overview 6
   1.5 What's in the box? 7
   1.6 Specifications 7
   1.7 Glossaries 8

2 KEY FEATURES 9
   2.1 Key features 9

3 INSTALLATION—GETTING STARTED 11
   3.1 Installation 11
   3.2 Getting started 12

4 USING YOUR .... INSTRUCTIONS – PROCEDURES 15
   4.1 Giving instructions and writing procedures 15
   4.2 Don't make assumptions 15
   4.3 Introduce procedures with a colon 16
   4.4 Put everything in chronological order 16
   4.5 Only have one instruction in each sentence 17
   4.6 Tell the reader about the expected results of each step 18
   4.7 Put a period at the end of each step 19
   4.8 Refer to buttons concisely 19
   4.9 Tips 20

5 TROUBLESHOOTING 21
- 5.1 What is a troubleshooting section? 21
- 5.2 Describing the user's problems 21
- 5.3 Describing solutions to the user's problems 22
- 5.4 Tables 23
- 5.5 Avoid using lists 24

6 WARNINGS AND RECOMMENDATIONS 25
- 6.1 *warning* vs *recommendation* 25
- 6.2 Where to locate warnings 25
- 6.3 Use imperatives to express recommendations 26
- 6.4 Express warnings using the same format and terminology 27
- 6.5 Repeating the words such as *not* and *never* is good practice in warnings 29
- 6.6 Explain the consequences of ignoring a warning 29
- 6.7 Use *if* to explain a consequence 30

7 Updates - warranty - contact details 31
- 7.1 Updates 31
- 7.2 Warranty 31
- 7.3 Contact details 32

PART II Writing Clearly, Concisely and Unambiguously

8 WRITING FROM A READER PERSPECTIVE 35
- 8.1 Address the reader directly 35
- 8.2 Only address the user using *you* when it is really necessary 36
- 8.3 Say what your product or service does, not what it is designed to do 37
- 8.4 Tell the users what they can do and how to do it 37
- 8.5 Avoid *should* 38
- 8.6 Be careful using words like *usually, normally* and *generally* 39
- 8.7 Make it clear to the reader whether he / she has to do something or whether the system does something automatically 39

9 AVOIDING REDUNDANCY AND LONG SENTENCES 41
- 9.1 Be concise 41
- 9.2 Use verbs rather than nouns 42
- 9.3 Use adjectives rather than nouns 42
- 9.4 Use verbs rather than adjectives that end in *-able* 43
- 9.5 Limit yourself to one idea per sentence 43

9.6	Avoid long sentences	44
9.7	Avoid parenthetical phrases	45

10 WORD ORDER ... 47
10.1	Key rules for word order	47
10.2	Subject	48
10.3	Verbs	49
10.4	Direct objects and indirect objects	51
10.5	Noun + noun sequences	52
10.6	Adjectives	54
10.7	*figure, table, appendix* etc.	55
10.8	Past participles	56
10.9	Adverbs of frequency + *also, only, just, already*	57
10.10	Adverbs of probability	58
10.11	Adverbs of manner	58
10.12	Adverbs that indicate a chronological order	59
10.13	Adverbs of time	59
10.14	Adverbs with more than one meaning	60
10.15	Negations	61

11 TERMINOLOGY ... 63
11.1	Refer to the same type of reader using the same term	63
11.2	Use the same terminology for the same scenario	64
11.3	Use the most specific word possible	66
11.4	Use the simplest word possible	66

12 AVOIDING AMBIGUITY ... 67
12.1	Check for ambiguous word order	67
12.2	*the former, the latter*	68
12.3	*if* and *when* clauses	68
12.4	Latin words and abbreviations	69
12.5	Be precise	69
12.6	*which*	70
12.7	*may might, can* and *will*	71

13 AUTOMATIC TRANSLATION ... 73
13.1	The advantages of automatic translation	73
13.2	Typical areas where Google Translate may make mistakes in English	74
13.3	How to improve the chances of getting an accurate automatic translation	77
13.4	Do not use Google Translate to check your English	79

PART III Layout and Order of Information

14 LAYOUT 83
   14.1   Decide which is clearer: one column or two columns 83
   14.2   Avoid long paragraphs 84
   14.3   Think about the best order in which to present
           information 85
   14.4   Put information in chronological order 86
   14.5   Set out the information in the simplest way 87
   14.6   Ensure grammatical consistency 88

15 HEADINGS 89
   15.1   Why use headings? 89
   15.2   Capitalization 89
   15.3   Follow a heading with some text 90
   15.4   Do not make headings part of the following text 90

16 PUNCTUATION 91
   16.1   Apostrophes (') 91
   16.2   Colons (:) 92
   16.3   Commas (,) 93
   16.4   Hyphens (-) 95
   16.5   Parentheses () 96
   16.6   Periods (.) 96
   16.7   Semicolons (;) 97
   16.8   Forward slash (/) 97

17 CAPITALIZATION 99
   17.1   Titles of documents 99
   17.2   Section headings 100
   17.3   Product names 100
   17.4   Days, months, countries, nationalities, natural
           languages 100
   17.5   Notes 101
   17.6   Acronyms 101
   17.7   OK 101
   17.8   Figures, tables, sections 102
   17.9   Steps, phases, stages 102
   17.10 Keywords 103

18 ABBREVIATIONS AND ACRONYMS 105
   18.1   Limit usage of abbreviations 105
   18.2   Quantities 105
   18.3   Introducing an acronym 105

	18.4	Punctuation	106
	18.5	Duplication	106

19 BULLETS ... 107
	19.1	Types of bullets	107
	19.2	When to use	108
	19.3	Punctuation	109
	19.4	Introducing bullets	109
	19.5	Avoid redundancy	110
	19.6	Bullets after section titles	110
	19.7	One idea per bullet	111
	19.8	Grammatical consistency	111

20 FIGURES, TABLES AND CAPTIONS ... 113
	20.1	Making reference to figures	113
	20.2	Numbering figures	114
	20.3	Abbreviations with figures, tables, appendices etc	114
	20.4	Captions to figures	115
	20.5	Use tables to show information quickly and clearly	115

21 DATES AND NUMBERS ... 117
	21.1	Day / Month / Year	117
	21.2	Decades	117
	21.3	Words (*twelve*) vs digits (12)	118
	21.4	Points vs commas	119
	21.5	Ranges, fractions, periods of time	119
	21.6	Percentages	120

22 GIVING EXAMPLES ... 121
	22.1	*for example*	121
	22.2	*e.g., i.e.*	122
	22.3	*etc*	122
	22.4	Dots (…)	122

23 REFERENCING ... 123
	23.1	Sections and documents	123
	23.2	Figures, tables, windows	124
	23.3	*the following*	124
	23.4	*above mentioned / as mentioned above*	125
	23.5	*hereafter*	126

24 SPELLING ... 127
	24.1	US vs GB spelling	127
	24.2	Technical words	127
	24.3	Misspellings that automatic spell checkers do not find	128

PART IV Typical Mistakes

25 COMPARISONS ... 131
25.1 Comparative vs superlative ... 131
25.2 Adverbs and prepositions used with comparisons ... 132
25.3 *the more... the more* ... 132

26 DEFINITE ARTICLE (*THE*), INDEFINITE ARTICLE (*A, AN*), *ONE* ... 133
26.1 Definite article: main uses ... 133
26.2 Definite article: other uses ... 134
26.3 *a / an* + singular countable nouns ... 135
26.4 Uncountable nouns ... 136
26.5 *a* vs *an* ... 137
26.6 *a / an* vs *one* ... 137

27 GENITIVE ... 139
27.1 General usage ... 139
27.2 Companies ... 141
27.3 Countries and towns ... 142
27.4 Periods of time ... 142

28 INFINITIVE VS GERUND ... 143
28.1 Infinitive ... 143
28.2 Gerund vs infinitive ... 143
28.3 Gerund ... 144
28.4 *by* vs *thus* + gerund ... 145

29 NEGATIONS ... 147
29.1 Position ... 147
29.2 Contractions ... 147
29.3 *no one* vs *anyone* ... 148
29.4 Double negatives ... 148
29.5 *also, both* ... 149

30 PASSIVE VS ACTIVE ... 151
30.1 Addressing the user ... 151
30.2 Referring to lists, figures, tables and documents ... 152
30.3 When the passive must be used ... 152

31 PRONOUNS ... 153
31.1 *you* ... 153
31.2 *we, us, our* ... 153
31.3 *he, she, they* ... 154
31.4 *users* ... 155

31.5	*it, this*	156
31.6	*one, ones*	156
31.7	*that, which, who*	157
31.8	*that* vs *which*	158

32 VOCABULARY 159

32.1	*allow, enable, permit, let*	159
32.2	*function, functionality, feature*	161
32.3	*(the) last, (the) next*	162
32.4	*login* vs *log in, startup* vs *start up*, etc.	162
32.5	*and*	163
32.6	*as, as it*	164
32.7	*both … and, either … or*	165
32.8	*even though, even if*	166
32.9	*in case, if*	166
32.10	*instead, on the other hand, whereas, on the contrary*	167

THE AUTHOR 169

Index 171

Part I
Structure and Content of a Manual

Chapters 1–7 describe the sections typically found in a user manual. For your particular product or service, you may not need all the sections. However, it is worth reading all these chapters as they contain useful tips for presenting information in a clear and user-friendly format.

1 TITLE—TABLE OF CONTENTS—ABOUT—INTRODUCTION—PRODUCT OVERVIEW—WHAT'S IN THE BOX

1.1 Title

Give your document a clear name.

Have a look through any user manuals that you have available and compare the various titles. Also note the:

- layout
- prominence given to the name of the product and service
- fonts, font style and font size
- color
- images
- logo

You will notice that there is a massive variety, and that the simplest form is generally the most effective. Typical titles include variations of the following:

Owner's Manual

User Handbook

Instructions on installation and use

Using your *Name of Product*

The last one is perhaps the most effective. To learn about the use of initial capitals in titles and headings see 17.1 and 17.2.

1.2 Table of Contents

This should appear on the first page of the manual, i.e. on the inside of the front cover. If the user is likely to refer to the table of contents on a frequent basis, then consider having it printed on the inside cover as an extendable flap that the user can open and see at the same time as consulting the rest of the manual.

The contents can be laid out in various ways. Here is a clear example showing the first and last items of a table of contents:

CONTENTS

Welcome to digital satellite TV! 1
      Getting started 3
      Your remote control 4
      Turning your digibox on and off 5
      Changing channels 6

- - - - - -

Connecting your digibox 28
      System set up 29
Further help 30
Glossary 31
Troubleshooting 32
After sales service 37
Specifications 38
Index 39

Note how:
- the main section titles are concise but clear in meaning
- the main section titles are in bold
- the subsections are indented and in normal script
- there is white space between dots and the titles and the page numbers

The above features make the Table of Contents much easier to access.

1.3 About

The 'About' section typically informs users of:

- the product name
- the product version
- the names of any other documentation they need to have read or be familiar with in order to read your doc. List such documentation under the heading 'Related Documentation'
- any prior technical knowledge that they will need in order to be able to understand your doc
- the meaning of any terminology / vocabulary that might not be familiar to them

1.4 Introduction / Product overview

In your introduction you can:

- thank the user for choosing your product (but this is not necessary)
- give a very brief overview of the product

Ensure you write from the reader's pint of view. In the example below bold is used to show the main differences between the two versions.

Yes	No
Thank you for choosing SHUB®, the sync & charge station that **acts** as a SuperSpeed USB 3.0 HUB and as a smart charging box. With HUB IT® **you can connect your computer** to as many as seven electronic devices, such as smartphones,…	Congratulations for choosing SHUB®, the **highly innovative** sync & charge station **that is designed to provide the capabilities** of a SuperSpeed USB 3.0 HUB, whilst simultaneously functioning as a smart charging box. **Due to the state-of-the-art technology employed in the design of** HUB IT®, **a computer can be connected** to as many as seven electronic devices, such as smartphones,…
SHUB® automatically **recognizes any devices that you plug in** and then charges them at the maximum speed possible. LEDs **tell you** the charging status of each device.	Any devices that **are plugged in are automatically recognized** by SHUB® and the **appliance** then charges them at the maximum speed possible. **Users are informed** of the charging status of each device **through the utilization of a set of** LEDs.

Note how in the 'Yes' version:

1. the descriptions are in the active (*you can connect, you plug in*) rather than the passive (*can be connected, are plugged in*)
2. the reader is addressed directly (*you*)
3. the most concise form is used (*LEDs*) rather than using long phrases full of redundancy (*through the utilization of a set of LEDs.*)
4. there are no subjective expressions (e.g. *congratulations, highly innovative, state-of-the-art technology*). Such expressions are more suitable for sales literature

Points 1–3 are basic rules for writing manuals and will be mentioned frequently throughout this book.

1.5 What's in the box?

The *What's in the box?* section tells users what should be contained in the box. This means that if something is missing, they can immediately contact your company.

Keep the list as clean as possible.

Yes	No
• SHUB®	• 1 x SHUB®
• Four removable modules with connectors	• 4 x removable modules with connectors
• 5V DC, 4A adapter	• 1 × 5V DC, 4A adapter
• Instruction manual	• 1 x Instruction manual
• Quick start guide	• 1 x Quick start guide

In any case, if you decide to use the multiplication symbol (x), then ensure it is not attached to the preceding number as this could confuse users. Also consider using written numbers (e.g. three) rather than digits (e.g. 3).

Yes	No
2 x USB 3.0 micro cables	
Two USB 3.0 micro cables	2x USB 3.0 micro cables

1.6 Specifications

Most manuals contain details of the specifications of the product. These are likely to be of interest to technical people rather then end users.

People buying household objects such as irons, tumble dryers, digiboxes and video / recording equipment are probably not particularly interested in immediately learning the dimensions and weight of the product they have just bought, or the audio frequency range, or the beam divergence of the laser.

If such specifications are not likely to be of great interest to the user, then you can locate this section at the end of the user manual rather than at the beginning.

1.7 Glossaries

Glossaries are lists of technical and semi-technical words that appear in the manual. For example, a Blue Ray Disc player might contain a glossary of such terms as: aspect ratio, bit rate, chapter, MP3, pulse code modification, parental control. Like the Specifications, most users will find the glossary of no interest and / or of little utility, so it is probably best located at the end of the document.

Below is a good example of how to write a glossary. It explains the terms often used in guides to software applications.

Check box: This is a box where the user can flag something.

Combo box: A text box with a list of available values attached to it. For example, the font size control in Word / Outlook / Excel, and so on. It is sometimes used to mean the same thing as "drop-down list". This distinction between "combo box" and "drop down list" is sometimes clarified with terms such as "non-editable combo box".

Diagram: A visual representation of a system, a technical concept, data flow, or suchlike.

Dialog box: A window where you can perform tasks. This is also known simply as a window. It is called a 'dialog' box because the computer uses it to 'tell / ask' you something or you use it to 'tell' the computer something.

Drop-down list: This allows you to choose one value from a list. When a drop-down list is inactive it displays a single value. When activated it displays (drops down) a list of values, from which the user may select one. When the user selects a new value the control reverts to its inactive state, displaying the selected value. A drop-down list differs from a combo box in that the entry portion of a drop-down list cannot be edited. The navigation field of a web browser is an example of a combo box rather than a drop-down list.

Figure: Any picture, diagram, or screenshot. It is the most commonly used word to refer to any kind of picture in a manual.

Window: An enclosed, rectangular area of a user interface where you can run a program, display data, and so on.

Note the style of the above glossary:

- key word in bold
- key word followed by a colon
- first word after the colon has an initial capital letter (e.g. *Any picture, On a, An enclosed*)
- user addressed directly (*you*)

2 KEY FEATURES

2.1 Key features

If you have not already described the key features of your product in the Introduction, then you could have a separate section. This gives users a quick preview of what your product or service does and how it will be useful for them. Below are extracts from some literature explaining some of the services that a bank offers.

EASY TO BANK WITH US

At XXX Bank, we want it to be easy for you to contact us and access your account.

Online Banking

Our secure online banking service provides access to a wide range of services at any hour of the day or night from any location. You can make payments, check your balance.

Telephone Banking

We offer a range of services including payments, checking balances and ordering cards. We have consultants on hand to answer your call Monday to Friday.

ATMs

Use your card at any cash machine to withdraw cash, pay in cash or cheques, get a mini-statement and check your balance.

The above extracts highlight that:

- this section does not have to be simply called 'key features' - you can think of a more meaningful heading
- before listing your key features you can have a one line-introduction, which summarizes the overall essence of the features
- the list of key features does not have to be in bullet form. Instead it can be a series of mini headings
- each mini section begins with a different grammatical form. When using bullets (see Chapter 19) it is a good idea to use the same grammatical form at the beginning of each bullet (e.g. *We aim to:* • *make you feel welcome...* • *handle your accounts properly....* • *put things right as soon as possible...* • *always understand your financial needs*). But when you have a series of mini sections, each of which is a few lines long, then it is not necessary to be consistent in the grammatical form

- the bank refers to itself using the first person pronoun (*we, our, us*). This makes them seem more 'human' and thus more likely to provide a friendly efficient service. Likewise, the client is referred to as *you*. See 8.1, 8.2 and 31.1 on the use of personal forms
- the helpdesk operators are referred to as *consultants* - this makes them sound like banking experts and is designed to give the reader more confidence in the service offered

3 INSTALLATION—GETTING STARTED

3.1 Installation

Before they can be used, many products (both hardware and software) need to be installed. It is critical to give users the correct procedure to follow to carry out installation. For details on how to write procedures see Chapter 4.

Below is an example from a manual on a superfast multiple charger. Note that the numbers in brackets refer to items in an illustration.

Installation SHUB® is a Plug & Play device, so installation is quick and easy.

1. Connect the power adapter supplied to the DC jack (1). The adapter is optional, but recommended. Without power, the system may not be able to feed the devices properly, as the current from the computer is very low.

2. Connect the micro USB port (2) to a USB 3.0 port of your computer using the USB 3.0 cable provided. The computer's operating system will automatically detect SHUB®.

    Notes:

- You can connect other devices to SHUB®, which will also be recognized by your computer.

- For each USB 3.0 port and each USB 3.0 slot there is an amber LED which lights when your device is charging. If a LED fails to illuminate this does not indicate a malfunction of the device, but only that the absorption of current is very low or zero.

In the above example note how:

- there is an introductory statement telling the user that installation is 'quick and easy'. Many non technical people (technophobes) can become anxious when installing an unfamiliar product. So the introductory statement is designed to reassure and relax such users

- the sentences are short and easy to follow

- the two steps in the installation process are introduced with numbered bullets. This helps to distinguish them from the two unnumbered bullets (in the Notes) which just give extra information

- the numbers in the bullets are the same as the numbers on the related diagram (not shown here)

- the first step gives an explanation of the importance of using an adapter, and the second step tells the user about the expected outcome after the second step has been completed (see 4.6 to learn why this is important).

3.2 Getting started

With many products you may not need to install any software. However, you may need to check a few things before you can actually use the product. The section which tells users about any preparations they need to make is generally called 'Getting started.' Here is an example from a digibox.

> GETTING STARTED This section gives you the information you need to start watching 3D digital satellite TV.
>
> For more detailed information about your digibox, use the Contents or Index to find the relevant section of this guide.
>
> --
>
> Before starting, make sure your Viewing Card is inserted the right way up in the slot marked Viewing Card on the front of your digibox. If you do not have a Viewing Card, call your broadcaster's helpdesk. For your broadcaster's helpdesk number, select the Telephone Numbers option from the Services screen.
>
> So that you can watch all the channels and services you want, you must leave your Viewing Card in your digibox at all times.

Note how:

- the first sentence describes what the section covers, i.e. it covers the essential things that users need in order to be able to carry out basic functions so that they can watch TV

- the second sentence refers to a different concept than in the first sentence, so it begins on a separate line. It tells users where they can find more detailed information about more complex features

- there is a line break between the first two sentences (which are an introduction to the section) and the following paragraphs which describe the preliminary preparations. The idea is to use white space, paragraphs and lines (dotted, continuous etc) to aid readers in their navigation of your document

- the term 'viewing card' is written in two different formats i) Viewing Card and ii) *Viewing Card*. The first is used to refer to the physical card. The second is in italics and refers to a location on the digibox. It functions as an adjective. The terms *Telephone Numbers* and *Services* are used in the same way, i.e. before an adjective. This simple convention helps readers to distinguish between the object itself and a function connected with the object

3.2 Getting started (cont.)

- the user is given details about what to do if they don't have a viewing card, what number they should ring, and where they can find this number. Providing this information gives the reader a positive experience—they are not frustrated by not having all the details they need
- the word *must* is used to describe obligatory requirements. *must* is unambiguous, it is clear to the reader that there are no other options

4 USING YOUR .... INSTRUCTIONS – PROCEDURES

4.1 Giving instructions and writing procedures

The most critical sections in a manual are those in which the reader is expected to follow a series of instructions (steps) in order to be able to carry out a task.

Writing procedures involves:

- identifying the major tasks and separating them into subtasks
- writing a series of steps that walk the user through each subtask (often presented as a list of bullets)
- not having two steps in the same sentence or bullet
- only giving the reader information at the exact moment that they need it (this is known as 'just in time' information)

Remember that users may be in a state of frustration as a result of having failed to carry out the task by themselves. So when they read your instructions, they need to be able to follow each step easily and clearly.

4.2 Don't make assumptions

Be very careful when making assumptions about what the user already knows and the level of their expertise. You are very familiar with the product or service that you are writing about. So you may forget to include certain steps. You may think that these steps are very obvious, but to many readers nothing will be obvious.

4.3 Introduce procedures with a colon

When you introduce a procedure, simply use a colon (:).

Yes	No
To install the software:	To install the software, do the following:
1 blah	1 blah
2 blah	2 blah

4.4 Put everything in chronological order

If you were writing about how to defuse a bomb, it would not be a good idea to say: *Cut the green wire having first ensured that the red wire has been disconnected.* Instead, you need to say:

1. Ensure the red wire is disconnected.
2. Cut the green wire.

4.5 Only have one instruction in each sentence

The following sentence is difficult to absorb because it contains a series of instructions in one long sentence.

Removing a Favorite channel

To remove a channel from your list of favorites, highlight it on the *Favorite Channels* screen, then press the *Favorite* (yellow) button, the tick will disappear indicating that the channel has been removed from your list of favorites.

Revised Version 1: three short sentences in one paragraph

Removing a Favorite channel

To remove a channel from your list of favorites, highlight it on the *Favorite Channels* screen. Then press the *Favorite* (yellow) button. The tick will disappear. The channel is now no longer in your list of favorites.

Revised Version 2: numbered bullets

Removing a Favorite channel

To remove a channel from your list of favorites:

1. Highlight the channel on the *Favorite Channels* screen.
2. Press the *Favorite* (yellow) button.

Outcome: The tick disappears. The channel is no longer in your list of favorites.

Revised Version 1 contains three short sentences, which are easy to follow. This solution uses a paragraph rather than a numbered list of instructions. It is suitable for when there are few instructions to follow, and for when the instructions themselves are quite intuitive and easy to apply.

Revised Version 2 is more appropriate for longer lists of instructions and when space is not an issue (e.g. on an online document).

4.6 Tell the reader about the expected results of each step

Users need to know the expected outcome in order to understand whether they have followed the instructions correctly. They do not want to be left in suspense!

When following instructions for a software application, the user needs to know what the screen will show after he/ she has performed a task. Below is an example from a software manual.

1. From the Main Menu, open the window showing the data you want to export.

2. Select *Current Page* to export all the data available for the window the command has been recalled from. Alternatively, select *All Result Rows* to export only the data retrieved via the search launched.

    Outcome: The File Download dialog window opens.

3. Click *Open* to open the Excel file. Alternatively, click *Save* and choose where you want to save the document.

    Outcome: Your Excel file appears in the location you have chosen.

In the above example note how the lines that describe the 'Outcome' are not numbered, i.e. they are not part of the numbered list. This helps them to stand out and also avoids confusion.

The example below is <u>not</u> good, because points 3 and 5 appear to be steps that the user has to follow, whereas in reality they simply show the outcome of the previous step/s.

1. From the Main Menu, open the window showing the data you want to export.
2. Select *Current Page* to export all the data available for the window the command has been recalled from. Alternatively, select *All Result Rows* to export only the data retrieved via the search launched.
3. The File Download dialog window opens.
4. Click *Open* to open the Excel file. Alternatively, click *Save* and choose where you want to save the document.
5. Your Excel file appears in the location you have chosen.

4.7 Put a period at the end of each step

It needs to be clear to the reader where one step ends and the next begins. This should be clear from the use of numbered bullets. However, it also helps to put a period (.) at the end of each step. This makes it clear that the step is ended.

4.8 Refer to buttons concisely

Imagine this situation. You are giving users of your software application a procedure to follow. This procedure involves the user clicking on buttons. You have provided a figure that clearly shows the buttons.

In the above situation:

- there is no need to use the word 'button'
- do not use 'button' with 'OK'. 'OK' is always a button and there can be no confusion for the user
- there is no need to put 'on' after 'click'
- it is not usually necessary to specify whether it is a left-hand click or a right-hand click. Note: The verb is 'to click' not 'to make a click'

	Yes	No
1	Click *Save As* and choose where you want to save the file.	Click on the *Save As* button and choose where you want to save the file.
2	Click *OK*.	Click the *OK* button.
3	Click *OK*.	Click on *OK*.
4	Click *Save As*.	Make a left-hand click on *Save As*.

4.9 Tips

A tip is a short sentence that informs the user about:

- a simple way of achieving what might appear to be a difficult task
- a typical problem that can be avoided

Let's imagine you are writing instructions on how to send an attachment via email. A typical procedure might end with the following steps:

5. Write your message.
6. Press *attachment* to upload the document to be attached. When the document has uploaded, the document icon will appear in the *attachments panel*.
7. Repeat Step 6 if you have other documents to send.
8. Check that all the documents you wish to upload are shown in the *attachments panel*.
9. Press *send*.

   Tip: When you want to send an attachment, it is easy to forget to upload the attachment. To avoid this problem, carry out Steps 6-8 before Step 5.

The tip in the example above is basically a warning or recommendation (see Chapter 6). It is not essential for users to follow the tip. In such a situation, terms like *warning, caution* or *attention* would be too strong.

Below is an another good example of a tip:

Installing modules in slots

Remove the SHUB lid by lightly pulling on the inner corners of the upper cover and lifting the cover (Fig. 2).

TIP: Only two sides allow the cover to be opened. If you pull up the corners slightly and the cover does not open, try opening from the other side. This means you can work out in advance which sides enable the cover to be opened.

5 TROUBLESHOOTING

5.1 What is a troubleshooting section?

A troubleshooting section lists the typical problems that a user may encounter, and provides solutions.

From your company's point of view, the troubleshooting section must be as clear and as comprehensive as possible so that your helpdesk will only have to deal with very special cases.

In all cases, you should try to predict the typical questions / scenarios that users are likely to encounter, and offer a clear practical solution.

Troubleshooting is also written *trouble shooting*, or alternatively you can use the heading *Solving Problems*.

5.2 Describing the user's problems

Below are some headings that describe four user problems. They are taken from a manual for a washing machine and are clear and effective. The use of bold is mine.

My machine **makes** a noise or **vibrates** in a spin program.

The program **takes** a long time.

My machine **will not start.**

The dispenser **will not close** properly.

Note the use of:

- *my* - this makes the manual seem more user-friendly
- the present tense to express a scenario using an affirmative verb
- *will* to express a scenario using a negation

5.3 Describing solutions to the user's problems

Let's imagine the same situation as in 5.2, i.e. problems with a washing machine. The solution can be presented in a series of instructions and questions (possibly in the format of a flow chart):

My machine will not start.

Close the door. Choose a program and then press 'ON'.

> Does the 'indicator' light come on after two seconds?
> Is the machine plugged in and switched on?
> Is the socket OK? Test with another appliance to check.
> Is the fuse in the plug OK?

Plug the machine in and turn the socket switch on.

If possible use another socket for the machine.

If not, replace it, see STEP 1: Electricity Supply

5.4 Tables

The example below comes from a manual for a digibox for receiving a satellite signal on a television.

It is clear, simple and effective.

On-screen messages

Message	Possible reason	What to do now
Please insert your Viewing Card.	There is no viewing card in the 'Viewing Card' slot in your digibox.	Insert your viewing card into the 'Viewing Card' slot.
You have entered your PIN incorrectly three times. PIN is now blocked for 10 minutes.	Your PIN has been entered incorrectly three times in a row.	You will not be able to access anything that needs a PIN for 10 minutes. If you have forgotten your PIN, call your broadcaster's helpdesk. To retrieve your broadcaster's helpdesk number, select the *Telephone Numbers* option on the *Services* screen.

General problems

Problem	Possible reason	What to do now
You can't find the remote control.		Use the buttons on the front of your digibox. You can perform most digibox functions using these buttons.

The above example highlights that:

- you can divide your troubleshooting section into subsections (in this case: *On-screen messages* and *General problems*)
- a clear layout and clear headings make it easy for readers to find what they want and resolve their problem
- your instructions should be exhaustive. In this case, the manual does not simply say call your broadcaster's helpdesk, but also where to find the telephone number for this helpdesk
- you do not need to fill in every cell (in this case, there is no need to explain the reason for the disappearance of the remote control!)

5.5 Avoid using lists

The following example is taken from a manual to a Blue Ray Disc (BRD) player. This example highlights how a list is NOT an effective method of presenting a troubleshooting section. For example in the third case (THE REMOTE CONTROL DOES NOT OPERATE CORRECTLY), there are several sequences of steps with no indication as to where each sequence begins.

I HAVE AN ON SCREEN MESSAGE.

You have attempted to carry out a function, which is not possible at this time.

Please follow the instructions as given on the display.

THE DISPLAY IS NOT ILLUMINATED

The AC mains lead has become disconnected.

Connect mains lead securely.

THE REMOTE CONTROL DOES NOT OPERATE CORRECTLY

The remote signal cannot reach your machine.

Point the remote at the front of the machine and make sure there are no objects in the way.

The remote is too far from the machine.

Move closer to the machine or replace the batteries.

The batteries have been inserted incorrectly.

Insert the batteries correctly.

The wrong remote control mode has been selected.

Switch over to BRD.

6 WARNINGS AND RECOMMENDATIONS

6.1 *warning* vs *recommendation*

A warning describes an action that the manufacturer considers will be dangerous, damaging or harmful either to the user, to the product or to whatever is used in conjunction with the product. The manufacturer is saying: *You must not do x.*

A recommendation is not as strong as a warning. The manufacturer is saying: *You can do x if you want. But we don't think it is a good idea.*

To protect your company from possible legal issues with clients, you need to be clear whether you are writing a warning or a recommendation.

6.2 Where to locate warnings

Preferably have a separate section for warnings, and locate it near the beginning of the document where there is perhaps a greater chance that the user will read it.

If you have to include a warning in the middle of another section, make sure that it stands out from the rest of the text. Ways to do this are to use:

- a warning symbol (such as a road-sign warning, i.e. an exclamation mark inside a triangle; an icon of a bomb; a hand pointing)
- bold, underlining and capital letters (but ensure you are consistent in their use)
- a different color
- a larger font
- a box around the warning text
- more white space than usual around the warning text

6.3 Use imperatives to express recommendations

The following examples come from some washing machine instructions. The use of bold is mine.

> **Empty** all objects from pockets, as they may damage the clothes and the machine.
>
> **Place** small items in a wash bag.
>
> Always **follow** the care label on the items when choosing the wash program.
>
> **Wash** non-colorfast objects separately as they may affect other items.

Note how in some cases the recommendations also describe the consequence of ignoring them. This helps the reader to understand whether to follow the recommendations or not.

6.4 Express warnings using the same format and terminology

There are several ways to give warnings. The example below is again taken from the instructions to a washing machine (the formatting is as in the manufacturer's instructions).

> Do not overload the machine (maximum load 5.5 kg). In addition to reducing the quality of the wash, this may also damage your laundry and the machine.
>
> We strongly recommend that you **do not** wash underwired bras in this machine. Should the wires become detached it could cause damage to your clothes and the machine.

The first instruction is very clear because it begins with *Do not*. This is 100% clear to the reader, who can have no doubt that a load greater than 5.5 kg represents a risk. A less common alternative is:

> You must not overload the machine (maximum load 5.5 kg).

The second instruction (*We strongly recommend*) is designed to protect the manufacturer from complaints by customers. The manufacturer knows that in any case people typically wash underwired bras in a washing machine rather than by hand. However, by writing explicitly that they 'strongly recommend' not using a washing machine for this purpose, they protect themselves. They could, of course, have simply written:

> Do not wash underwired bras in this machine.

In fact, if all warnings are given in the same way they are easier to read and comprehend, as highlighted in the Yes version below. Whereas in the No version, the warnings are given in three different ways.

YES	NO
Do not overload the machine.	**Do not** overload the machine.
Do not wash underwired bras in this machine.	**We strongly recommend that you do not** wash underwired bras in this machine.
Do not wash white with colored items.	**It is advisable not** to wash white with colored items.

Now let's analyse a bad example. It comes from from the instructions to a charger for a powered golf trolley.

It highlights that:

- numbering bullets is not effective when listing random warnings
- beginning each sentence with a different grammatical form makes the instructions more difficult to absorb quickly

6.4 Express warnings using the same format and terminology (cont.)

IMPORTANT SAFETY INSTRUCTIONS

1. Before using the charger, read all the instructions and cautionary markings on the charger.
2. Do not expose the charger to rain or snow.
3. Always stand the charger on a hard service in a well ventilated area.
4. To reduce the risk of damage to the electric plug and cord, pull by the plug rather than the cord when disconnecting from the mains.

A better version would be:

IMPORTANT SAFETY INSTRUCTIONS

> Your charger has been designed with your safety in mind. However, before using the charger, read all the instructions and cautionary markings on the charger. In addition you MUST take the following precautions:

- **Never** expose the charger to rain or snow.
- **Never** pull by the plug rather than the cord when disconnecting from the mains. Correct disconnection will reduce the risk of damage to the electric plug and cord.
- **Always** stand the charger on a hard service in a well ventilated area.

The revised version:

- begins with an introductory remark
- lists the precautions in the same way (all beginning with either *never* or *always*)
- uses bold (*never*, *always*) and capital letters (*must*) to give emphasis

6.5 Repeating the words such as *not* and *never* is good practice in warnings

In most of the manual, your aim is to write as concisely as possible. But when you give the user warnings, being concise is not the best approach.

The use of *Do not* at the beginning of each sentence is more effective than using *Do not* once to introduce all the warnings. The example below is again from a charger for a powered golf trolley.

Repeating the words *Do not* combined with the use of bold underlines the importance of the warning.

YES	NO
Do not expose the charger to rain or snow.	Do not:
Do not disassemble the charger.	• Expose the charger to rain or snow.
Do not operate the charger with a damaged chord or plug.	• Disassemble the charger.
	• Operate the charger with a damaged chord or plug.

A user who happens to glance at the page where the 'No' version above appears, might not even notice the words *Do not* and simply read: *Disassemble the charger*

6.6 Explain the consequences of ignoring a warning

It is good policy to explain the consequences of the user's performing something that is not recommended. Avoid using *may, might* and *could* indifferently. They generally have an identical meaning, but the reader may think there are differences between them. Instead, when talking about possible consequences, use

- *may* for something that is possible but not certain
- *will* for something that is certain

Here is an example from a portable music device. The use of italics is mine.

AVOIDING HEARING DAMAGE

Do not wear the headphones for prolonged periods as this *may* cause irritation or pain to your ears.

Do not set the volume to maximum for more than five minutes at a time. Prolonged and frequent listening at high volume *will* damage your hearing.

6.7 Use *if* to explain a consequence

As a general rule, always use the shortest, simplest and least ambiguous grammatical form.

YES	NO
Do not wash underwired bras in this machine. **If the wires** become detached, this may damage your clothes and the machine.	Do not wash underwired bras in this machine. **Should the wires** become detached it could cause damage to your clothes and the machine.

if only has one meaning. On the other hand, *should* is used in many contexts, and its use when talking about conditions and consequences might confuse the reader.

7 UPDATES - WARRANTY - CONTACT DETAILS

7.1 Updates

To 'update' a manual means to add and delete information in order to reflect any new features or enhancements that have been made to your product / service.

Typically, a manual states what version it is (or the product / service version) and when it was updated. If the new version requires users to update their operating systems, then such a requirement should be included.

7.2 Warranty

A warranty is a written guarantee that your company will give to users / purchasers, promising to repair or replace it if necessary within a specified period of time. Warranties usually have exceptions that limit the conditions your company will be obligated to respect in order to rectify a problem.

Take legal advice before writing this section. Ensure that you take all precautions to draft the warranty in a way that protects your company as much as possible. Do not simply cut and paste the warranty section from documents that you find on the Internet.

7.3 Contact details

Make sure you regularly update your contact details (names of departments, addresses, helpdesk phonelines etc).

Explain clearly what each contact is for.

Only provide phone numbers if you really want to be contacted by telephone. Lay out phone numbers in groups of between two and four digits - this makes the number easier to understand.

Providing names of specific people can be risky, as they may leave the company.

>**Helpdesk services**: helpdesk@company.com, +39 050 788 0045 876
>
>**Licence agreements:** licence@company.com
>
>**Sales:** sales@company.com
>
>**Customizations and enhancements**: custom@company.com

Part II
Writing Clearly, Concisely and Unambiguously

John Ruskin, English art critic and social thinker said that:

It's far more difficult to be simple than complicated.

Chapters 8–12 outline key rule for writing documents from a reader's perspective. By following these rules, your readers should find reading your manual a pleasurable and non-frustrating experience. As Bruce M Cooper, author of "Writing Technical Reports", wrote:

Human beings are not logical mechanisms into which information can be fed.

Part 2 ends with Chapter 13 on how to use automatic translation tools. This chapter should be useful if you have documentation in your language that you want to translate into English.

8 WRITING FROM A READER PERSPECTIVE

8.1 Address the reader directly

Manuals are generally intended for one type of reader – the user. Try to interact directly with the reader by using:

1. *you* instead of *the user* or *the reader* (but see 8.2 below)
2. *you* instead of an impersonal or abstract form
3. the imperative instead of the gerund or other grammatical forms

When you use *you*, occasionally the resulting sentence is longer. This is not a problem, the document will still be more readable.

For cases of documents where there might be more than one possible type of reader see Chapter 11 Terminology.

	YES	NO
1	With filters **you** can focus on the records **you** are interested in.	The use of filters enables **the user** to focus on the **desired set** of records.
1	If **you** are familiar with the classic XYZ gateway …	If **the reader** is familiar with the classic XYZ gateway …
2	When the translation has been completed, the Send button is enabled. This **allows you** to submit the translation to..	When the translation has been completed, the Send button is then enabled **to allow** the translation to be submitted to…
3	If you want to use a filter from a column header, **click** on the arrow.	Filters are also available on each column header **by clicking** on the arrow in the column header.

8.2 Only address the user using you when it is really necessary

If you use *you* too often the manual will begin to sound like an email or a conversation. Do NOT use *you* in the following cases, when you are:

1. instructing the reader what to do. Use the imperative instead
2. giving the reader a procedure to follow. Use the imperative and bullets instead
3. telling the reader how to do something. Use the infinitive instead
4. referring the reader to a particular point in the manual
5. talking about a series of different types of users

	YES	NO
1	**Use** the pages in the Configuration folder to set the automatic system behaviors and interface, as detailed in the subsections below.	The pages available in the Configuration folder **enable you** to set the automatic system behaviors and interface, as detailed in the subsections below.
2	**Select the instrument**, the correct date and time, and the upper and lower boundary for the price. **Click** on "Update".	**You must select** the instrument, the correct date and time, the upper and lower boundary for the price and then **you have to click** on "Update".
3	**To open** the XYZ Trans table: Click on Plug-ins → Zeta → XYZ Trans	By clicking on Plug-ins → Zeta → XYZ Trade **you can open** the XYZ Trans table.
4	**Below is** the list of values that can be selected for each drop-down control.	Below **you** can find the list of values that can be selected for each drop-down control.
4	See *Configuring the system* on page 21.	If **you** need further information, see *Configuring the system* on page 21.
5	The users that can connect to the Control Panel are listed in the Users page of the Settings section. **They** can have one or more roles that define **their** permissions. On the basis of **their** role, **they** can access certain windows and features.	The users that can connect to the Control Panel are listed in the Users page of the Settings section. **They** can have one or more roles that define **their** permissions. On the basis of **your** role, **you** can access certain windows and features.

8.3 Say what your product or service does, not what it is designed to do

If you write a sentence such as *KwikTrans is designed to produce accurate translations*, the reader cannot be sure if the term *design* is meant to imply an initial objective or whether in reality KwikTrans does produce accurate translations.

So don't create doubt in the user's mind. Avoid expressions such as the following:

> was designed to
>
> was intended to
>
> was aimed at
>
> has the following aims:

Instead write:

> KwikTrans produces accurate translations.
>
> This service guarantees better results.
>
> This machine is 30 % faster.
>
> This service has the following features:

Similarly, instead of writing *This document aims to describe the main features of ...*, simply write: *This document describes the ...*

8.4 Tell the users what they can do and how to do it

Rather than describing the features and functions of your product or service, tell your readers how they can use the product / service.

YES	NO
With KwikTrans you can: • see alternative translations for certain phrases • instantly check whether the spelling is US or GB • insert comments, footnotes and captions	In addition to providing an automatic translation into the language of interest, the KwikTrans application provides alternative translations for certain phrases, indicates whether the spelling is US or GB, and also provides the flexibility for the insertion of comments, footnotes and captions.

8.5 Avoid *should*

Do not put doubt in your reader's mind. In the example below the reader is not sure whether the start up window really will be displayed, and whether the list will appear.

> To launch the KwikTrans application, click the KwikTrans icon on the desktop. This **should** display the KwikTrans startup window, where you **should** be able to see a list of choices regarding.

Instead tell readers exactly what will happen next, using the present tense or *will*:

> This **displays** the KwikTrans startup window, where you **will** see a list of choices regarding.

Alternatively:

> The KwikTrans startup window **is displayed**, and **shows** a list of choices regarding.

So, always provide feedback for each action so that the users can see whether the action has worked correctly.

The modal verb *should* implies that there is some kind of option involved, i.e. that a recommendation is being made that does not necessarily have to be followed. In the example below, it is not clear if the action must be performed or is simply a good idea.

> Before installing the software, users **should** verify that their operating system is compatible.

Better versions are as follows:

> **Ensure** that your operating system is compatible, before installing the software. Failure to do so will damage your hard disk. Below is a list of compatible systems:

> **You must ensure** that your operating system is …

If you are informing users of an obligation (rather than recommending) then:

- use the imperative (e.g. *verify* rather than *should verify*) or *must*
- use the strongest word possible (e.g. *ensure* rather than *verify*)
- explain the consequences of an action not being taken (e.g. *Failure to do so will…*)
- explain what to do next

For more on recommendations and warnings see Chapter 6.

8.6 Be careful using words like *usually, normally* and *generally*

In the example below, the reader may be worried about what *usually* (in the first line) implies.

1. Open the language drop-down menu. This usually displays 10 languages from which you can choose the one of interest.
2. Select the language/s of interest.
3. Press Translate.

The reader will automatically assume that there is a possibility that something will go wrong, and that 10 languages may not appear.

If your application risks not doing what it designed to do, then you need to inform readers:

> Open the language drop-down menu. This usually displays 10 languages from which you can choose the one of interest.
>
> Note: If less than 10 languages are shown, and your chosen language is not on the list, then open the Settings menu and …

Otherwise, simply remove the word *usually*.

8.7 Make it clear to the reader whether he / she has to do something or whether the system does something automatically

Readers need to know who does what. Do they have to do something, or are other people (e.g. the systems administrator, a technician) or systems involved?

Such confusion and ambiguity often result from the use of the passive form (see Chapter 30), as highlighted in the No example below.

YES	NO
When **you select** a language **for display** in the Languages window, **KwikTrans takes** the default values for spelling, accents and hyphenation from the language's details.	When a language **is selected and displayed** in the Languages window, the default values for spelling, accents and hyphenation **are taken** from the language's details.

9 AVOIDING REDUNDANCY AND LONG SENTENCES

9.1 Be concise

Don't use:

1. meaningless abstract words
2. meaningless descriptive words
3. unnecessary introductory phrases
4. unnecessary link words
5. references to earlier unspecified parts of the document (examples: *as mentioned above, as already stated*). Remember that the reader will not read your document starting at page 1 and finishing at page 100

	YES	NO
1	This supports the installation.	This supports the **activity** of installation.
1	Achieving this is difficult.	Achieving this is a difficult **task**.
1	We believe the results are significant.	We believe the results are of significant **value**.
2	They should be green and round.	They should be green **in color** and round **in shape**.
3	Note that the sum of the values needs to be lower than....	**It is worth noting / Bear in mind** that the sum of the values
4	This component does not support XYZ.	**Furthermore / In addition / In particular / It is worth noting that** this component does not support XYZ.
5	Market data are not required by the system.	**As stated above**, market data are not required by the system.

9.2 Use verbs rather than nouns

Use:

1. a verb rather than a noun
2. one verb rather than a noun + verb

YES	NO
This was used to **calculate** the values.	This was used **in the calculation of** the values.
This allows you **to transfer** the money. This allows the money **to be transferred**.	This allows **the transfer** of the money to be performed.
Brazil **was compared** to England.	A **comparison was made** between Brazil and England.
Brazil **performed** much better than England.	Brazil **showed a** much better **performance** than England.

9.3 Use adjectives rather than nouns

Prefer:

1. verb + adjective constructions to verb + noun
2. comparatives with adjectives rather than nouns

YES	NO
This method has quite an **efficient** calculation process. Calculations with this method are quite **efficient**.	This method shows quite a good **efficiency** in the calculation process.
X is **more homogeneous** than Y.	X has **a higher homogeneity** with respect to Y.

9.4 Use verbs rather than adjectives that end in -able

When telling users:

1. what they can do, use you *can* + verb instead of an adjective ending in *-able*
2. how they can do something, use the imperative instead of an adjective ending in *-able*

YES	NO
1	**You can customize / configure** the user interface.	The user interface is **customizable / configurable**.
2	**Download** the key from our website.	The key is **downloadable** from our website.

9.5 Limit yourself to one idea per sentence

Ideally, each sentence should contain only one piece of new information and should be no more than about 20 words long.

Break the sentence into two if:

1. to include two pieces of information in one sentence requires more than 20 words
2. the two parts of the original sentence are not strictly connected, but the second is, for example, simply a consequence of the first

YES	NO
1	The X can be configured using the Preferences menu of the XX **window**. Its main components	The X can be configured through the Preferences menu of the XX **window,** and its main components are the RFQ window and the RFQ blotter. (25 words)
2	When the trade has been completely allocated, the Send button is **enabled**. This allows you to submit the selected allocation breakdown to the market.	When the trade has been completely allocated the Send button is then **enabled** to allow the actual sending of the selected allocation to the market. (25 words)
2	The residual quantity is updated in real **time**. The field is colored according to the value displayed.	The residual quantity is updated in real **time**, and is colored differently according to the value displayed. (17 words, two different concepts)

9.6 Avoid long sentences

In long sentences:

1. replace the **which** clause by beginning a new sentence. Note: repeating the same word (in this case *divisions*) is not bad style..
2. start a new sentence when there is a link word or other adverb / preposition, e.g. and, then, when, moreover, in addition,
3. use bullets where possible

	YES	NO
1	Our company has many R&D divisions where innovative research **is carried out. These divisions** are located in various parts of …	Our company has many R&D divisions where innovative research **is carried out, which** are located in various parts of …
2	This value specifies the maximum period the gateway can remain in a waiting status while establishing the connection to the **servers**. The gateway **then** stops the connecting process and tries later.	This value specifies the maximum period the gateway can remain in a waiting status while establishing the connection to the **servers; then** the gateway stops the connecting process and will try later.
2	**When** the trading connection is running again, **if** the XX is set to 1, the gateway **then** tries to authenticate all the traders. **If** the trader is successfully authenticated, the X field of the corresponding Y record is to "Running". **If not**, the CstatusTradingStr field remains "Disconnected".	Then, **if** the XX is set to 1, **when** the trading connection is running again, the gateway tries to authenticate all the traders: **if** the trader is successfully authenticated, the X field of the corresponding Y record is to "Running"; **otherwise**, the CstatusTradingStr field remains "Disconnected".
3	For example, users can: • create customer groups and product groups. • define rules • add restrictions • create teams.	For example, users can create customer groups and product groups, define rules, add restrictions, and create teams.

9.7 Avoid parenthetical phrases

Subjects often get separated from their verbs by parenthetical information contained between two commas or in brackets. In such cases, the use of commas and brackets breaks the flow of the sentence and makes it harder to understand immediately. To avoid this problem:

1. rearrange the sentence so that the subject and verb are next to each other. The order you use will depend on the emphasis you want to give
2. when the parenthetical information is rather long, split the sentence up

	YES	NO
1	**This feature will** only be of limited use, owing to its high cost. *Or:* Owing to its high cost, **this feature will** only be of limited use.	**This feature**, owing to its high cost, **will** only be of limited use. *Or:* **This feature** (owing to its high cost) **will** only be of limited use.
1	The **vegetables were cooked** in the oven **and then** served with the main course.	The **vegetables**, cooked in the oven, **were served** with the main course. The **vegetables,** which had been cooked in the oven, **were served** with the main course.
2	**We believe** the results are significant given their innovative nature. When they are analysed they should help in our understanding of the diffusion of this virus in the world today.	The analysis of the results, **which we believe are of a significant value given their innovative nature**, should help in the understanding of the diffusion of this virus in the world today.

10 WORD ORDER

10.1 Key rules for word order

Put the most important information at the beginning of the sentence. The most important information is generally new information or negative information.

English word order is strict: 1) subject 2) verb 3) direct object 4) indirect object – all these four elements should go as close as possible to each other.

Adjectives go before nouns.

Past participles generally go after nouns.

Most adverbs go immediately before main verb.

10.2 Subject

The subject generally contains the most important information.

1. Put the subject as near as possible to the beginning of the sentence
2. In affirmative phrases put the subject before the verb
3. Avoid using an impersonal **it** at the beginning of the sentence. Instead use modal verbs (*might, need, should* etc) or an adverb

	Yes	No
1	**Several techniques** can be used to address this problem.	To address this problem **several techniques** can be used.
1	**Time and cost** are among the factors that influence the choice of parameters.	Among the factors that influence the choice of parameters are **time and cost**.
1	**The old system** should **thus** not be used.	**For this reason**, it is not a good idea to use the **old system**.
2	The new **software** has arrived.	It has arrived the new **software**.
2	The **method** is important.	It is important the **method**.
3	Users **should** be distributed evenly.	**It is recommended** to distribute users evenly.
3	You **can** do this with the new system. This **can** be done with the new system.	**It is possible** do this with the new system.
3	This **is probably** be due to poor performance.	**It is possible** that this is due to poor performance

10.3 Verbs

Make sure the verb is near the beginning of the sentence and next to the subject. If the subject is incredibly long, the reader will be left waiting to know what the verb is. To avoid this problem:

1. use an active verb

2. shift the verb to the beginning of the sentence. This may involve changing the verb

3. divide up the long sentence into two shorter sentences (see 9.6)

	Yes	No
1	**ABC generally employs** people with a high rate of ...	People with a high rate of intelligence, an unusual ability to resolve problems, a passion for computers, along with good communication skills **are generally employed** by ABC.
2	This data shows that **there are** significant correlations between ...	This data shows that significant correlations between the cost and the time, the time and the energy required, and the cost and the age of the system **exist.**
3	People with a high rate of intelligence **are generally employed** by ABC. They must also have other skills including: an unusual ability to ...	People with a high rate of intelligence, an unusual ability to resolve problems, a passion for computers, along with good communication skills **are generally employed** by ABC.

Sometimes the verb has multiple subjects. In such cases:

1. use bullets

2. shift the verb immediately after the first subject. In this case, the first subject is generally the most important

3. if there is an adverb of manner (such as *easily, quickly*) then transform this adverb into an adjective, and shift the verb to the beginning

10.3 Verbs (cont.)

Yes	No
1	The following can be configured: • fonts • filter functionalities	Fonts, filter functionalities, blotters, and message bars **can easily be configured**.
2	Fonts can be easily configured **as well as** filter functionalities, blotters …	Fonts, filter functionalities, blotters, and message bars **can easily be configured**.
3	It is **easy** to configure fonts **as well as** filter functionalities, blotters …	Fonts, filter functionalities, blotters, and message bars **are easily configurable**.

10.4 Direct objects and indirect objects

Generally English prefers this order: 1) direct object 2) indirect object:

1. when a verb is followed by two possible objects, place the direct object (i.e. the thing given or received) before the indirect object (the thing that the direct object is given to or received by). This kind of construction is often found with verbs followed by *to* and *with*

    Examples: *associate X with Y, apply X to Y, attribute X to Y, consign X to Y, give X to Y (or give Y X), introduce X to Y, send X to Y (or send Y X)*

2. if the direct object is very long and consists of a series of items, you can put the **indirect object** after the first item and then use 'along with'
3. as an alternative to 2, you can use a colon to introduce a list
4. as an alternative to 2, you can use bullets

	Yes	No
1	We can **associate a high cost** with these values.	We can **associate** with these values **a high cost**.
2	We can associate **a high cost with these values**, along with higher overheads, a significant increase in man hours and several other problems.	We can associate **with these values a high cost**, higher overheads, a significant increase in man hours and several other problems
3	We can associate several factors **with these values**: a high cost, higher overheads,…	
4	The following can be associated **with these values**: • a high cost • higher overheads • a significant increase in man hours	

10.5 Noun + noun sequences

Do not randomly use a string of nouns. There are no clear rules regarding this area of English grammar. Below are some guidelines:

1. when referring to components, platforms, products, services etc, put the descriptive word first (i.e. the type of something) and then the generic word. So, we say the *edit menu* and not the *menu edit* because we are describing the type of menu, so *edit* acts like an adjective and thus comes before the noun it describes

2. be careful when using the genitive (see Chapter 27). If in doubt, say *the main features of KwikTrans* and not *KwikTrans's main features*

3. generally speaking use the singular for the descriptive word. This means you should say *client requirements* and not *clients requirements*, even if there are several *clients* involved

There are many exceptions. So if you think you have an exception (e.g. *workspaces management, solutions provider*), check it with Google first.

	Yes	No
1	Click on the **Edit menu**.	Click on the **menu Edit**.
2	**Sales of KwikTrans** have exceeded expectations.	**KwikTrans's sales** have exceeded expectations.
3	Submitting translations	Translation submitting Translations submitting Translations' submitting

10.5 Noun + noun sequences (cont.)

You cannot indiscriminately put nouns in front of each other. It makes the sentence hard to understand and often leads to ambiguity.

To avoid misunderstanding on the part of your reader, use a preposition (*of, for, by*), and where necessary convert the nouns into verbs:

1. *of* = which belongs to
2. *for* = for the purpose of
3. *by* = how something is done

	Yes	No
1	The streets **of / in** Rio.	Rio streets.
2	Instructions **for** boiling potatoes.	Potatoes boiling instructions.
3	Quantifying surface damage **by measuring** the mechanical strength of silicon wafers.	Silicon wafer mechanical strength measurement for surface damage quantification.

Using nouns as adjectives may be wrong in certain cases, but not in others. Unfortunately as far as I know there are no rules.

In any case, strings of nouns and adjectives must be used if they are names of pieces of equipment or methods. For example:

A recently developed reverse Monte Carlo quantification method

A Hitachi S3500N environmental scanning electron microscope

10.6 Adjectives

Below are some general rules for the use of adjectives:

1. adjectives generally come before the noun that they describe
2. an exception to Rule 1 is *available*, which generally comes after
3. if you have to put an adjective after the noun, then it should be preceded by a relative clause (*which* + verb)
4. you cannot put an adjective between two nouns
5. do not put an adjective before a noun that it does not describe

	Yes	No
1	These are **new products**.	These are **products new**.
2	The **products available** at the moment are:	The **available products** at the moment are:
3	This product, **which is new** to our range, is available in three different versions.	This product, **new** to our range, is available in three different versions.
4	The **main** interface of the editor	The editor **main** interface
4	The **computational** complexity of the algorithm.	The algorithm **computational** complexity
5	The **main contribution** of the document.	The **main document** contribution.

Note:

Comparative adjectives (see 25.1, 25.2) behave like other adjectives:

1. they go before the noun they describe
2. if they must appear after the noun, then precede them with *that*

	Yes	No
1	This solution has **more serious** drawbacks than the other solution.	This solution has drawbacks **more serious** than the other solution.
2	The application returns only the **results that are the most relevant**.	The application returns only the **results most relevant**.

10.7 *figure, table, appendix* etc

The rules below refer not just to *table*, but similar words such as *figure, appendix, diagram,* and *screenshot*:

1. to avoid the use of the passive, put the word *table* at the beginning of the sentence
2. only put the subject before *table*, if the subject is short
3. if the subject is a long series of items put *table* first
4. alternatively to Rule 3, put *table* after the first item

	Yes	No
1	Table 1 highlights the most significant values.	In Table 1 are highlighted **the most significant values**.
2	The most significant values are highlighted in Table 1.	In Table 1 are highlighted **the most significant values**.
3	**Table 1 highlights** the most significant values along with blah blah blah blah blah blah blah blah blah blah blah blah blah blah blah blah blah blah.	The most significant values along with blah blah blah blah blah blah blah blah blah blah blah blah blah blah blah blah blah blah **are highlighted in Table 1**.
4	The most significant values **are highlighted in** Table 1 **along** with blah blah blah blah blah blah blah blah blah blah blah blah blah blah blah blah blah blah	The most significant values along with blah blah blah blah blah blah blah blah blah blah blah blah blah blah blah blah blah blah **are highlighted in Table 1**.

10.8 Past participles

A past participle is a word like *found, managed, shown*. Past participles are located as follows:

1. in 99% of cases past participles can always go after the noun, and in 50% they cannot go before. So, put them after!

2. don't follow a past participle with a description, unless this description is introduced by *which* (things) or *who* (people)

3. a better solution for Rule 2 is to use two sentences instead of one

	Yes	No
1	It shows details of all the results **found**.	It shows details of all the **found** results.
1	The data **reported** show that...	The **reported** data show that...
2	The data, **which are reported** in the table below, are very useful for...	The data, **reported** in the table below, are very useful for...
3	The data **are reported** in the table below. They are very useful for...	The data, **reported** in the table below, are very useful for...

Notes

In some cases both positions are possible. The second example below shows that the past participle after the noun (*actions specified*) indicates that further details will be given.

> It shows details of all the **specified actions**.
>
> It shows details of all the **actions** (which are) **specified** (in the manual).

Be careful with **used**. Before the noun it means 'second hand', after the noun it means 'which is used'

> I bought a **used car**.
>
> This was the **application used** by the testers.

10.9 Adverbs of frequency + *also, only, just, already*

Adverbs of frequency (e.g. *always, sometimes, occasionally*) and words like *also, just, already,* and *only*, are usually placed:

1. immediately before the main verb
2. immediately before the second auxiliary when there are two auxiliaries
3. after the present and past tenses of *to be*
4. you can put some adverbs (*sometimes, occasionally, often, normally, usually*) at the beginning of a sentence, if you want to give special emphasis

	Yes	No
1	You **only / also / just** need to sign the document.	You need **only / also / just** to sign the document.
1	We don't **usually** go to Corsica on holiday.	We **usually** don't go to Corsica on holiday.
2	We would **never** have seen him otherwise.	We **never** would have seen him otherwise.
2	This may not **always** have been the case.	This may not have been **always** the case.
3	The supplier is **always** on time with deliveries.	The supplier is on time **always** with deliveries. The supplier **always** is on time with deliveries. **Always** the supplier is on time with deliveries.
4	**Normally** X is used to do Y, but **occasionally** it can be used to do Z.	

Note:

When *only* is associated with a noun rather than a verb, it is located before the noun:

> **Only** Emma has been to Japan. (Not anyone else)
>
> Emma has **only** been to Japan. (Not anywhere else)

10.10 Adverbs of probability

Adverbs of probability (e.g. *probably, certainly, definitely*) go immediately before the:

1. main verb

2. negation (*not* and contractions e.g. *don't, won't, hasn't*)

Yes	No
1 She will **certainly** come.	She **certainly** will come. She will not come **certainly**.
2 She will **probably not** come.	She **probably** will **not** come. She will not **probably** come. She will **not** come **probably**.
2 She **definitely hasn't** read it.	She **hasn't definitely** read it.

10.11 Adverbs of manner

An adverb of manner describes how something is done or to what extent e.g. *quickly, completely*. Some adverbs of manner can go before the verb. But, since all adverbs of manner can always also go after the verb or noun, it is best to put them there.

1. subject + verb + adverb of manner + full stop (.)

2. subject + verb + noun + adverb [+ rest of phrase]

Yes	No
1 This program could help **considerably**.	This program could **considerably** help.
2 This program will help system administrators **considerably**. This program will help system administrators **considerably** to do x, y and z.	This program will help **considerably** system administrators. This program will help **considerably** system administrators to do x, y and z.

10.12 Adverbs that indicate a chronological order

When you are listing events:

1. put the adverb (*firstly, secondly* etc) at the beginning of the phrase. You can say *firstly* or *first*, *secondly* or *second*, etc
2. *then* can go at the beginning of the sentence or before the main verb

	Yes	No
1	**First / Firstly**, we will do X. Then we will do Y. **Finally**, we will do Z.	We will **firstly** do X. Then we will do Y. We will **finally** do Z.
2	Initially, we used X. **Then** we decided to use Y. At the beginning we used X, we **then** decided to use Y.	

10.13 Adverbs of time

Adverbs of time:

1. usually go at the end of the phrase, particularly if they consist of more than one word
2. when used in contrast with each other, they go at the end
3. in some cases (e.g. *today, tomorrow, tomorrow evening*) they can go at the beginning for emphasis

	Yes	No
1	We will go there once or twice a week / as soon as possible.	Once or twice a week / as soon as possible we will go there.
1	We will go there **immediately**.	We will **immediately** go there. We will go **immediately** there.
2	We will go there tomorrow morning not tomorrow evening.	Tomorrow morning we will go there not tomorrow evening.
3	**Today**, we are going to talk about the position of adverbs.	We **today** are going to talk about the position of adverbs.

10.14 Adverbs with more than one meaning

There are a few adverbs that change meaning depending on whether they are located before or after the verb:

normally: before = usually, after = the opposite of abnormally

clearly: before = obviously, after = without difficulty

fairly: before = quite, after = in the right proportion

10.15 Negations

Negations generally contain key information. For example, they tell readers what they must not do, what features are not available or do not function. This means putting the following near the main verb:

1. *not* and *no*

2. *only, rarely, seldom, never* and other similar adverbs that contain negative information

Yes	No
1	I **don't** think this is a good idea.	I think this is **not** a good idea.
1	This **didn't** seem to be the case.	This seemed **not** to be the case.
1	There is **almost no** documentation on this particular matter.	Documentation on this particular matter **is almost completely lacking**.
2	This **rarely** happens when the user is online.	The number of times this happens when the user is online is generally **very few**. The frequency of this event when the user is online is **rare**.
2	We **only** realized this at the end of the tests.	We realized this **only** at the end of the tests.

Notes:

1. if you put *only* or a negation (e.g. *never, nothing*) or negative adverb of frequency (*rarely, seldom*) as the first word of a phrase, then you must invert subject and object as if you were forming a question

2. point 1 is difficult to remember, so it is best to use the normal word order instead

Yes	No
1	**Rarely does this happen** when the user is online.	**Rarely this happens** when the user is online.
1	… **nor is** the correlation between X and Y significant.	… also the correlation between X and Y **is not** significant.
2	This **rarely** happens when the user is online.	
2	The correlation between X and Y is **not** significant **either**.	

11 TERMINOLOGY

11.1 Refer to the same type of reader using the same term

Some documents require clear clarification regarding which types of users can perform certain functions.

you: when you are referring directly to the reader, and the reader is also the person who is using the product or service

user: when the reader is not necessarily the person who is going to use the product or service

authorized user: when you need to distinguish between 'normal' users and those with particular permissions

technician, systems administrator etc: when you need to specify that the user is a specific technical person

11.2 Use the same terminology for the same scenario

Below is an extract from a section entitled 'Safety' in a manual for the user of an electric kettle (a kitchen item used to boil water).

It highlights the dangers of using several different terms to refer to exactly the same idea. It is a bad example.

SAFETY

Make sure the kettle is switched off before lifting or pouring.

Warning: Do not open the lid while the water is boiling.

Make sure the lid is secure before switching the kettle on.

Caution: Do not operate the kettle on an inclined surface.

before plugging in

Make sure your electricity supply is the same as the one shown on the underside of your kettle.

Important – UK only

The appliance must be protected by a 13A approved (BS1362) fuse.

WARNING: this appliance must be earthed

The problems with the above safety instructions are:

- the reader does not know if there is a difference between *warning, caution, Important*, and *WARNING*. Is one more serious than another? Or do they in fact mean the same thing? And why are some instructions preceded by such warning words whereas other instructions begin with *make sure* (is it more serious to open the lid while boiling than not to switch off the kettle before lifting?)
- is there a difference between *the kettle* and *the appliance*?
- does the final warning (i.e. earthing the appliance) refer to the UK or to the entire world?

Moreover, the layout is confusing, there seems to be no consistency, and there is no logic in the order that the instructions are presented. A better version is:

11.2 Use the same terminology for the same scenario (cont.)

SAFETY

Please read these instructions carefully before using your kettle.

* Before plugging in your kettle for the first time, ensure it is earthed.

Note: If you are in the UK, your kettle must be protected by a 13A approved (BS1362) fuse.

* Make sure your electricity supply is the same as the one shown on the underside of your kettle.
* Do not operate the kettle on an inclined surface.
* Make sure the lid is secure before switching the kettle on.
* Do not open the lid while the water is boiling.
* Make sure the kettle is switched off before lifting or pouring.

The revised version is better because:

- the instructions are in a logical order. First the things to do before you start using the kettle are listed, and then the things to be careful of while using the kettle.
- given that each instruction is practically equal in importance, there is no use of bold, capital letters or underlining.

The above version could be improved by dividing into subsections, as shown below:

SAFETY – please read carefully

Before you plug in your kettle for the first time you MUST make sure that:

1) your electricity supply is the same as the one shown on the underside of your kettle.
2) the kettle is earthed
3) for UK users only: the kettle is protected by a 13A approved (BS1362) fuse

Please read the following warnings carefully before using your kettle.

- Do not operate the kettle on an inclined surface.
- Make sure the lid is secure before switching the kettle on.
- Do not open the lid while the water is boiling.
- Always switch the kettle OFF before lifting or pouring.

Note how numbered bullets in the first part of the safety instructions above attract attention as they give the idea that certain steps must be followed.

11.3 Use the most specific word possible

If there is a recognized technical word for a certain action, then use it. In the following example, *open* has been used in all four sentences. It is only appropriate in the first. In the other three sentences the more appropriate verb is in square brackets.

Open the CD tray.

Open the setup.exe file. [run]

The Print dialog **opens**. [is displayed]

The application opens automatically. [starts]

11.4 Use the simplest word possible

There are glossaries of Simplified Technical English (STE). These glossaries recommend using the most commonly used words to describe something. For example, the glossaries recommend using the 'simple' verbs in the first column in the table below, rather than the multi-syllable verbs in the second column.

YES (according to STE)	NO (according to STE)
show	demonstrate
help	facilitate
start	initiate
change	modify
stop, end	terminate

'Simple terms' is somewhat subjective. It means 'simple' for a native speaker. However, a non-native speaker may be more familiar with the multi-syllable words. In fact, multi-syllable words may be similar to words in their own language (e.g. French, Spanish, Italian and Greek), or may be what they have regularly seen in technical and scientific reports.

In any case, it does usually make sense to use shorter words when writing technical documents. They:

- occupy less space
- are quicker to read and absorb

12 AVOIDING AMBIGUITY

12.1 Check for ambiguous word order

Ambiguity arises when a phrase can be interpreted in more than one way due to the position of the words within the phrase.

> Technical writers should avoid annoying readers.

It is not clear if *annoying* describes the readers, or whether it refers to what writers should avoid doing. The two possible meanings are clarified below:

> Technical writers should avoid readers who are annoying.
>
> Technical writers should write in a way so that readers are not annoyed.

Below is another example where poor word order can create initial confusion:

> To obtain shades of red, palettes and other formatting devices can be used.

Readers of the above sentence may initially think that *red* and *palettes* are part of the same list. Readers will only understand that *palettes and other formatting devices* is the subject of the verb when they get to the end of the sentence. To avoid this problem you could write:

> You can use palettes and other formatting devices to obtain shades of red.

We tend to read words in small groups. Often we think that if two or three words immediately follow each other they must be related in some way. Again, the example below is confusing:

> The EU adopted various measures to combat these phenomena. This resulted in smog and pollution levels reduction.

When we read *resulted in smog and pollution*, our initial interpretation is that the smog and pollution are the result of the EU's measures. Then when we move on and read *levels* we have to reprocess the information. Instead, you want readers to understand everything immediately. A much clearer version is:

> The EU adopted various measures to combat this phenomena. This resulted in a reduction in the levels of smog and pollution.

For more details on word order see Chapter 10.

12.2 *the former, the latter*

Avoid using *the former* and *the latter* because it may not be clear which element you are referring to. Your our main aim is to make things clearer for the reader, so the best solution is always to repeat the key words, as in the examples below.

Yes	No
In this recipe we used potatoes, carrots and beans. This is common practice with this kind of cooking. **The beans** can, of course, be steamed.	In this recipe we used potatoes, carrots and beans. This is common practice with this kind of cooking. **The latter** can, of course, be steamed.
Such an unsolicited bandwidth request can be incremental or aggregate. If it is **aggregate**, the X indicates the whole connection backlog. Blah blah blah blah blah blah blah blah blah blah blah blah blah blah blah blah blah. Whereas if it is **incremental**, the X indicates the difference between its current backlog and the one carried by its last bandwidth request.	Such an unsolicited bandwidth request can be incremental or aggregate. In the **latter** case, the X indicates the whole connection backlog. Blah blah blah blah blah blah blah blah blah blah blah blah blah blah blah blah blah blah. In the **former** case, the X indicates the difference between its current backlog and the one carried by its last bandwidth request.

12.3 *if* and *when* clauses

Sentences beginning with *if* are often used to introduce a negative condition. For example,

> If you do not study, you will not pass the exam.

> If your warranty has expired, your product will not be protected.

This means that when you are referring to a positive situation, it may be better to introduce the situation with *when* instead of *if*.

The example below is from a manual for a product called SHUB. SHUB has two distinct functions: it is a SuperSpeed USB HUB and also a charging box. However, in the 'No' example below, it may seem that its function as a charging box may not be desired.

YES	NO
When SHUB is not connected to your computer, SHUB works as a high-performance smart charging box.	**If** you do not connect SHUB to your computer, the device will work as a high-performance smart charging box.
When SHUB is not connected to the computer, the device becomes a high-performance smart charging box.	**If** you connect devices to SHUB without the hub being connected to the computer, they will be charged optimally.

12.4 Latin words and abbreviations

Experts suggest that Latin words and abbreviations should be avoided because they may be unfamiliar to many people. There is certainly confusion between *e.g.* and *i.e.* (see 22.2) and *NB* can easily be replaced by Note:

Below are ways to avoid some common Latin expressions.

YES	NOT IN TECHNICAL DOCUMENTS, BUT NORMALLY OK
There are many ways to do this, **for example,** do x, do y, do z.	There are many ways to do this, **e.g.** do x, do y, do z.
There are two ways to do this, **that is,** x and y.	There are two ways to do this, **i.e.** x and y.
You can configure properties of the user interface, **for example** the color for the flashing of the cells, and the color for the **background**.	You can configure properties of the user interface: **the** color for the flashing of the cells, the color for the background **etc**.
A can send messages to B and **the other way round**. A can send messages to B, **and B to A**. A can send message to B. **Conversely**, B can send messages to A.	A can send messages to B and **vice versa**.
Note: It is dangerous to ...	**NB** It is dangerous to ...

12.5 Be precise

Where possible, avoid adjectives such as *large* or *small* and *rather* and *quite* as well as expressions such as *a certain amount*. Such words and expressions can be interpreted in many ways and can therefore be vague. Be as exact as you can.

YES	NO
You can translate up to **100 pages** of text within **10 s**. Note: KwikTrans has a **95%** **accuracy rate**. This means you will need to revise the resulting text.	**Large amounts** of text can be translated with KwikTrans within a **short timeframe**. Clearly there will be a **certain percentage** of errors within the resulting text, so users are advised to dedicate **some time** to checking the text.

12.6 *which*

which (see also 31.8) generally refers to the noun that it follows. In cases of possible ambiguity, avoid using *which*. Instead, split the sentence and repeat the subject. In the 'No' example below, the position of *which* initially seems to refer to Table 2. But in fact it refers to set of common rules.

Yes	No
Each language is characterized by a set of common rules, as reported in Table 2. **This set** highlights the structure of that particular language.	Each language is characterized by a set of common rules as reported in Table 2 **which** highlights the structure of that particular language.

When the *which* clause could refer to several but not all elements, remove *which* and repeat the specific elements. In the second 'No' example below, *which* could refer to A and B, B and C, or even A, B and C.

Yes	No
Examples of this kind of data are **Selected bonds, their Proportions** and the Sub Index Weightings. **Selected bonds and their Proportions** are normally established once a month.	Examples of this kind of data are Selected bonds, their Proportions and the Sub Index Weightings, **which are normally established once a month**.
Examples of this kind of data are A, B and C. **A and B** are normally established once a month.	Examples of this kind of data are A, B and C, **which** are normally established once a month.

12.7 *may might, can* and *will*

The differences between the modal verbs that express possibility are very subtle. Below are just some very general guidelines:

may, might: possibility, but not a certainty

can: ambiguous, could either be a possibility or a certainty

will: certainty

Below are some examples explaining the differences.

Before quitting the application, save any changes.

1) Failure to save changes **may** result in data being lost.
2) Failure to save changes **might** result in data being lost.
3) Failure to save changes **could** result in data being lost.
4) Failure to save changes **can** result in data being lost.
5) Failure to save changes **will** result in data being lost.

1-3 indicate that there is no certainty that data will be lost. You may be lucky, and your data will be saved. Within the context of a manual, there is no difference between *may*, *might* and *could*. However, *may* is the verb that is most used.

4 (*can*) indicates that this is a typical event, but there is no guarantee that it will happen.

5 (*will*) indicates an absolute certainty.

Here is another example to explain the difference between can and may:

It **can** rain a lot in England, especially in the winter.

Next week I am going to London, it **may** rain but I hope not.

The first example refers to a typical event – this is what generally happens in England. However, there could be exceptions, and in some winters it does not rain much.

The second example talks about a future event. It indicates a probability but not a certainty.

13 AUTOMATIC TRANSLATION

Throughout this chapter I will use the acronym GT to refer to Google Translate and automatic translation in general. Clearly, there are differences between for example, Bing and Google Translate, but for the purposes of this chapter such differences will be ignored.

For more on using Google Translate see Chapter 21 in *English for Academic Correspondence and Socializing* (Springer).

13.1 The advantages of automatic translation

In my experience as a translator I have found that automatic translation software, such as Bing and Google Translate (GT), is very useful for translating technical documents, particularly manuals. However, this accuracy strictly depends on

1. the language you are translating from – the greater the number of speakers of your language, the better the translation is likely to be
2. how much you modify the text of the original language before submitting it for translation

The second factor is crucial. Before submitting your text to GT, you need to make it more 'English' for example by:

- changing the word order to reflect English word order
- reducing the length of sentences
- replacing pronouns (e.g. *it, one, them*) with their respective nouns
- removing redundancy

From my own use and my analysis of my clients' translations, by using a text modified as suggested above, you are likely to produce a better translation with GT than if you started from scratch doing a manual translation. It will also take you considerably less time.

However, GT does make mistakes, so you MUST revise the translation or get a native speaker to edit and proofread it for you.

13.2 Typical areas where Google Translate may make mistakes in English

If you decide to translate your manual using automatic software, you need to be aware of the kinds of mistakes the software might make. Below I have listed the most common mistakes which, at the time of writing, GT makes.

WORD ORDER

GT's main difficulty is with word order, i.e. the position of nouns, verbs, adjectives, and adverbs. If in your language you put the verb before its subject, or if you put an indirect object before the direct object, then GT will not be able to create the correct English order (i.e. the reverse of the order in your language).

PLURAL ACRONYMS

In English, we say *one CD* but *two CDs*. Most other languages do not have a plural form for acronyms, and thus say *two CD*.

GT is able to recognize this for common acronyms such as CD, DVD and PC, but not for very technical acronyms.

TENSES

GT sometimes changes the tense from the original. For example, you may use the future tense and GT will translate it into the present tense. In some cases, GT may be correct. For example, if in your language you say 'I will tell him when I *will see* him', GT will correctly translate this as 'I will tell him when I *see* him'. This is because in a time clause, *when* takes the present and not the future. However, very occasionally GT makes mistakes when it changes tenses, so it is wise to check very carefully.

UNCOUNTABLE NOUNS

An uncountable noun (see 26.4) is a noun that cannot be made plural and which cannot be preceded by *a/an* or *one*. For example *information* is uncountable. This means you cannot say *an information, one information, two informations, several informations*.

The problem with uncountable nouns is that the surrounding words (i.e. articles, pronouns and verbs) must also be singular. This means that if, for example, *information* is countable in your language, GT will probably make errors with the surrounding words, as highlighted in this example (note: the example is NOT in correct English):

> These information are vital in order to understand xyz. In fact, they are so crucial that ...

13.2 Typical areas where Google Translate may make mistakes in English (cont.)

There are two solutions. You can modify your own language so that you put the surrounding words into the singular. Or you can check the English version. Whichever solution you use, the aim is to produce the following correct sentence:

> This information is vital in order to understand xyz. In fact, it is so crucial that ...

or:

> This information is vital in order to understand xyz. In fact, such information is so crucial that ...

See 26.4 for a list of common uncountable nouns.

VERY SPECIALIZED VOCABULARY

GT's dictionaries are huge but do not cover absolutely every word. If GT doesn't know a word, it will normally leave it in the original language.

WORDS WITH MORE THAN ONE MEANING

GT generally manages to guess the right meaning when translating into English because it looks at the surrounding words (i.e. how words are collocated together). In any case, you need to check carefully that GT has translated with the meaning you intended.

STRINGS OF WORDS USED IN COMPUTER TERMINOLOGY

If you use English phrases such as *status no-provider* in your own original language, sometimes GT will modify these when 'translating' and produce, for example, *provider-no status*. Essentially, you just need to check that any English in your source text has not been 'translated' by GT.

NAMES OF PEOPLE

At the time of writing, GT tends to translate people's first names and sometimes surnames. This should not be a problem in manuals as names of people do not usually appear. In any case, be aware that GT makes some rather unexpected translations.

ACCENTS AND SINGLE QUOTES

Does your native language use accents? If it does, then read on.

If you are, for example, French, then GT is helped considerably if you use the correct accents. Note how GT translates these two titles of a French medical paper in two ways depending on whether the accents are inserted or not. Interestingly, both translations would be possible, but one of the two might not reflect the author's real intention.

13.2 Typical areas where Google Translate may make mistakes in English (cont.)

> *Mesurer la qualité de vie: une nécessité en thérapeutique cancérologique*
>
> GT: Measuring quality of life: a need *for therapeutic oncology*
>
> *Mesurer la qualite de vie: une necessite en therapeutique cancerologique*
>
> GT: Measuring quality of life: a need in *oncology therapeutics*

Below is the same paper title, but this time in Italian. In this case, if the accents are correctly inserted in the original text, then GT provides the correct translation. Unlike with French, GT also provides exactly the same translation if the accents are not inserted at all.

> *Misurare la qualità della vita: una necessità per l'oncologia terapeutiche*
>
> *Misurare la qualita della vita: una necessita per l'oncologia terapeutiche*
>
> GT: Measuring the quality of life: a need for therapeutic oncology

But if the accent is placed after the final letter using a single quote (i.e. the ' character), which is a typical device used by Italians who don't have accents on their keyboards, GT gets confused and thinks a quotation is being given.

> *Misurare la qualita' della vita: una necessita' per l'oncologia terapeutiche*
>
> GT: Measure the quality 'of life: a necessity' for Therapeutic Oncology

Of course, words may change meaning depending on whether there is an accent or not. Here are two examples in French:

> *Le poisson est sale* = The fish is dirty.
>
> *Le poisson est salé* = The fish is salty.
>
> *Les moines aiment les jeûnes* = Monks like fasting.
>
> *Les moines aiment les jeunes* = Monks like young people.

So if your language has accents, you need to be aware that GT may produce unusual results!

SPELLING

GT uses US spelling. This is generally not a problem. But if your document requires UK spelling, then you will need to set your final spell check to UK spelling.

If words are misspelled in the original, then GT will either not translate them (if such a combination of letters does not exist – an English example would be *fomr*) or will mistranslate the word (e.g. *from* vs *form*).

13.3 How to improve the chances of getting an accurate automatic translation

The success level of a Google translation depends to a large extent on how similar the construction of your language is with respect to the normal structure of English.

One solution is to modify the version in your own language before you submit it to translation. Some of the most important modifications to make are listed below.

SYNTAX

Put the subject as near as possible to the beginning of the sentence and the main verb next to it. Put adjectives before their associated nouns. Make sure that the direct object precedes the indirect object. For the rules of English word order (see Chapter 10).

SENTENCE LENGTH AND PUNCTUATION

Limit the parts of your sentence to one or two (see 9.5 and 9.6).

Different languages use punctuation in different ways. Before you submit your text for translation, if possible try to punctuate it in an English way (see Chapter 16). Keep the sentences short, replace semi colons with full stops, and where appropriate use commas to break up the various parts of the sentence.

USE ACTIVE RATHER THAN PASSIVE SENTENCES

The advantage of an active sentence is that it must contain a subject, and this subject must precede the verb (in English). This means that GT is likely to produce a more accurate translation (see 30.1).

REPLACE ANY PRONOUNS WITH THE NOUNS THAT THEY REFER TO

Pronouns in English can be very ambiguous (see Chapters 12 and 31) because it may not be clear for the reader what they refer to. If you replace them with the noun they refer to, GT will make a more accurate translation. This is because Google works by looking for similar sequences of words in translations that it has already done. Words such as *it, they, them, one* can obviously be associated with many hundreds of thousands of other words. More concrete words such as *screen, mouse* and *modem* will be associated with fewer other words.

13.3 How to improve the chances of getting an accurate automatic translation (cont.)

DO NOT USE SYNONYMS FOR KEY WORDS

The more synonyms you use to express the same concept (see Chapter 11), the greater the chance that GT will make a mistake. Imagine for example that you are a doctor. In this case, a key word would probably be *patient*. Consequently you should always use *patient*, rather than finding synonyms such as *subject, participant, member, case, sufferer* etc. In the field of medicine, the term *patient* is more specific than the other synonyms. GT may link the other synonyms with non medical cases, and thus choose the wrong translation.

13.4 Do not use Google Translate to check your English

If you write something directly into English, you may think that you can use GT to check your English by translating it back into your own language to see if it makes sense. Unfortunately this does NOT work.

When you write in English you are naturally translating directly from your own language. So, if you submit your English text into GT and translate it back into your own language, the translated text in your own language will probably be very good because the structure of your English is based on the structure of your language.

However, this does NOT indicate that your English version is correct. It only indicates that the resulting text in your own language is what you wanted to say in English.

For example, let's imagine you have written a non grammatical sentence such as *I am here since yesterday*. You have used the simple present (*I am*) because in your language this is the tense you would use. In reality in such cases the correct tense in English is the present perfect, so the sentence should be *I have been here since yesterday*.

If you get GT to translate *I am here since yesterday* into your own language, then GT's translation will probably look correct in your language because it is a literal translation. However, although the translation into your own language is correct, the original English is not correct.

So, do not use GT to check the grammar of your English.

Part III
Layout and Order of Information

The following chapters provide some general guidelines on how to increase the readability of your manual by laying it out clearly and organizing the information in a logical way. The underlying rationale is to make the experience of the reader as pleasurable and as rapid as possible.

Before choosing a format for your manual, look at other manuals that you have available. The manuals do not have to be for the same type of product or service. In fact it is helpful to see a wide variety. Analyse the manuals, and decide which ones are laid out in the most effective way—you may decide to choose a combination of features from various manuals.

The best manuals (at least from a layout and visual point of view) that I looked at while preparing this book were for products by Apple, Google, Ikea, Microsoft, and Sky.

14 LAYOUT

14.1 Decide which is clearer: one column or two columns

You can write an extremely clear manual using either one column or two columns. What is important is the width of the column and the amount of white space.

One column which stretches from one side of the page to the other side is difficult to read. One column with plenty of white space in the left hand margin is easy to read. This is highlighted in the example below, which gives instructions on charging the battery from a golf trolley.

How to get the most from your battery

⚠ Always recharge your battery as soon as possible after use – certainly within 24 hours. Failure to carry this out can cause irreversible damage.

- Only use the charger provided with the golf trolley. Other chargers work at different rates and can severely shorten the battery's life.

- The battery should be used for a maximum of one round (i.e. 18 holes) between charges. It must be recharged even if fewer than 18 holes were played.

As highlighted by the above example, you can use the left hand margin for:

- making section headings stand out
- adding warning symbols, icons etc

You can also use two columns for A4 size pages. But ensure there is plenty of white space. Two columns in an A5 format looks very cramped and is hard to read.

14.2 Avoid long paragraphs

Below is a section entitled *Cleaning* taken from the instruction manual for a cooker hood (i.e. the device that houses an extractor fan, typically located above the gas or electric rings of a cooker). Version 1 has one long paragraph, which makes the information quite difficult to read and absorb.

VERSION 1 - ONE LONG PARAGRAPH

CLEANING

For your own safety and in the interests of hygiene your cooker hood needs to be kept clean. A build up of grease or fat from cooking could cause a fire hazard. Never use excessive amounts of water when cleaning, particularly around the control panel area. The metal casing and grille assembly should be cleaned at least once a month to keep it looking like new. Wipe over the hood with a soft cloth wrung out in hot water and containing a mild household cleaner and dry with a soft cloth. Always wear protective gloves when cleaning the hood. **Note: Never use scouring pads or abrasive cleaners as they might scratch or damage the surface.**

Version 2 highlights that using several paragraphs makes it easier for readers to read and absorb information.

VERSION 2 - SEVERAL SHORT PARAGRAPHS

CLEANING

For your own safety and in the interests of hygiene your cooker hood needs to be kept clean. A build up of grease or fat from cooking could cause a fire hazard.

Never use excessive amounts of water when cleaning, particularly around the control panel area.

The metal casing and grille assembly should be cleaned at least once a month to keep it looking like new. Wipe over the hood with a soft cloth wrung out in hot water and containing a mild household cleaner. Dry with a soft cloth.

Always wear protective gloves when cleaning the hood.

Note: Never use scouring pads or abrasive cleaners as they might scratch or damage the surface.

14.3 Think about the best order in which to present information

The example in version 2 in 14.2 is not in the best possible order. Below is the same text, but this time I have inserted numbers at the beginning of each sentence in order to explain in what order the info is presented.

(1) For your own safety and in the interests of hygiene your cooker hood needs to be kept clean. (2) A build up of grease or fat from cooking could cause a fire hazard.

(3) Never use excessive amounts of water when cleaning, particularly around the control panel area.

(4) The metal casing and grille assembly should be cleaned at least once a month to keep it looking like new. (5) Wipe over the hood with a soft cloth wrung out in hot water and containing a mild household cleaner. Dry with a soft cloth.

(6) Always wear protective gloves when cleaning the hood.

(7) Note: Never use scouring pads or abrasive cleaners as they might scratch or damage the surface.

The order in which the information is presented is as follows:

1) warning that hood needs to be kept clean (no specific reason for this safety warning is given)
2) reason for safety warning
3) caution about what materials to use when cleaning the hood
4) statement of how often particular parts of the hood need to be cleaned
5) instruction about how to clean the hood
6) caution about what to wear when cleaning the hood
7) caution about what materials to use when cleaning the hood

The above order (1-7) is not optimal. Version 3 (below) is a better solution.

VERSION 3 - INFO IN MOST LOGICAL ORDER: 1) WARNINGS 2) INSTRUCTIONS

CLEANING

Warnings

To avoid **fires** from a build up of grease or fat from cooking, keep your cooker hood clean.

When cleaning:

- **Always** wear protective gloves.
- **Never** use scouring pads or abrasive cleaners as they might scratch or damage the surface.
- **Never** use excessive amounts of water, particularly around the control panel area.

14.3 Think about the best order in which to present information (cont.)

When to clean the hood and how to clean it:

The metal casing and grille assembly should be cleaned at least once a month to keep it looking like new.

1) Wring out a soft cloth in hot water containing a mild household cleaner.
2) Wipe over the hood.
3) Dry with a soft cloth.

Version 3 first tells the reader about all the risks of (not) cleaning. It makes more sense to put warnings first, as readers need to know about what risks are involved before they learn how to carry out the cleaning itself. Note how bold is used to emphasize the importance of these warnings.

After the warnings, the reader is then given instructions on how to carry out the cleaning. These are now in short sentences and are presented as a series of steps. This type of presentation makes the steps easy to see on the page of the manual and thus easier to follow.

Version 3 requires more space on the page, but uses fewer words and is much clearer for the reader. It is always worth using more space when giving safety warnings (see Chapter 6).

14.4 Put information in chronological order

Your guide should be like a map showing the reader the direction to follow. Try to write in a step-by-step way, with each step logically following the previous one. This generally entails putting information in chronological order. So instead of saying:

The vegetables should be served with the main course after they have been **cooked.**

Say:

Cook the vegetables and then serve them with the main course.

14.5 Set out the information in the simplest way

Use a simple layout. This makes it much easier for the reader to immediately identify the important points. The example below highlights how NOT to lay out information.

> **3 Set up**
> **3.1 Installation**
> To run the gateway KwikTrans must already be installed on the user's computer. The version required is:
> 8.3.3 or later
> Also, in addition java runtime must be installed:
> 1.4.0_01

The example above:

- is visually cramped (there is little use of white space between lines)
- uses the passive and is thus not user friendly
- has a lot of redundancy (the phrase *must be installed* is used twice)
- does not make use of bullets
- looks totally unprofessional and will immediately create a bad impression

Below is a revised version. It is much more concise and clear. Note how it

- uses the minimum number of words (see Chapter 9)
- addresses the user directly using *you* (see 8.1 and 8.2) and uses the active form
- uses interlinear white space and bullets to help the reader to see and assimilate the information quickly
- clearly tells the user what the main section and subsections are about

> **3 Set up**
> This section describes the set up procedure.
> 3.1 Requirements
> To install the gateway you will need:
>
> - KwikTrans 8.3.3 or later
> - Java runtime 1.4.0_01

14.6 Ensure grammatical consistency

Within the same sentence or consecutive sentences, do not mix verbs and nouns when they refer to the same type of action or event. If possible, prefer verbs. In the example below, the writer users a noun (*acquisition*) and then a verb (*sending*).

X is used for the **acquisition** of Y and for **sending** Z.

The correct form is:

X is used to **acquire** Y and **send** X.

The same rules apply to bullets (see Chapter 19).

15 HEADINGS

Note: For the sake of clarity, all examples of headings in this chapter are in SMALL CAPS.

15.1 Why use headings?

Headings are useful for:

- helping users to navigate your document
- breaking up blocks of text

15.2 Capitalization

Only use an initial capital letter for the first word. Then put all the other words in lower case.

Do not use a full stop (.) at the end of a heading.

YES	NO
INSTALLING THE SOFTWARE	INSTALLING THE SOFTWARE.
The software requires version 2.1, or higher, of …	The software requires version 2.1, or higher, of …

15.3 Follow a heading with some text

Every heading should be followed by some text, even if just one sentence.

YES	NO
1 INSTALLING THE SOFTWARE **This section describes how to install the KwikTrans software.** 1.1 SYSTEM REQUIREMENTS:	1 INSTALLING THE SOFTWARE 1.1 SYSTEM REQUIREMENTS:

15.4 Do not make headings part of the following text

Section headings stand alone: they are not part of the text.

You cannot begin the first sentence that follows a heading with *it* or *this*.

it and *this* always refer back to something in the previous sentence. In the case of headings, there is no previous sentence.

YES	NO
2.3 ACCURACY OF THE COMPUTED SOLUTION **The accuracy** depends on the precision of the machine and…	2.3 ACCURACY OF THE COMPUTED SOLUTION **It** depends on the precision of the machine and…

16 PUNCTUATION

Punctuation shows the grammatical relationships between words, phrases, and sentences. It is also used to highlight particular words, and to show how concepts are connected together.

In technical writing it is best to use short, concise sentences so that the meaning becomes clearer.

As a test for clarity, try removing the punctuation from a sentence. If the resulting sentence does not make sense, it probably needs rephrasing, especially if it originally had a lot of commas.

For the use of initial capitalization, see Chapter 17.

Essential rules for punctuating sentences and paragraphs:

- use only commas (,), full stops (.) and colons (:)
- do not use need semicolons (;) or dashes (-)
- avoid parentheses ([ ]) and quotation marks (" ")
- avoid contractions such as *doesn't, isn't,* and *can't.* Instead, use *does not, is not,* and *cannot*

16.1 Apostrophes (')

Do NOT use an apostrophe to make:

1. acronyms and dates plural
2. contracted forms. Contracted forms are not generally used in technical documents. This is particularly true for negations, where the fact that a statement is negative is made much clearer by being written with *not* rather than *n't*

16.1 Apostrophes (') (cont.)

Yes	No
1 We bought six **PCs**.	We bought six **PC's**.
1 The company was founded in the 198**0s**.	The company was founded in the 198**0's**.
2 **Do not** use an apostrophe ... **Do NOT** use an apostrophe ...	**Don't** use an apostrophe ...
2 **Let us** assume that X = Y.	**Let's** assume that X = Y.

For the use of the apostrophe in genitives see 27.1.

16.2 Colons (:)

Use a colon to introduce:

- bullets and lists
- commands
- equations
- figures and tables

Yes	No
The figure below shows the quarterly yield:	The figure below shows the quarterly yield.
The values are listed below: • blah • blah • blah	The values are listed below. • Blah • Blah • Blah

Use a colon after *Note* and then begin the next word with a capital letter.

Yes	No
... using this function.	... using this function.
Note: The latest version of X must be installed for the function to work optimally.	**Note: the** latest version of X must be installed for the function to work optimally.

16.3 Commas (,)

Use commas to ensure clarity:

1. you shouldn't normally need more than two commas in a sentence. If you rearrange the sentence you may find that one comma is enough
2. to avoid long sentences (20 words or more – see Chapter 9), consider writing two sentences instead of using a comma
3. if the sentence contains a list of items, then you probably need to use bullets instead of using a series of commas
4. if you have a list of three items or more and decide not to use bullets, then use a comma before *and*. The comma highlights that the penultimate and last element are separate items

	Yes	No
1	If you install X after you have installed Y, this may cause damage. *one comma*	If you install X after, rather than before, installing Y, then this may cause damage. *three commas*
		Damage may be caused if you install X after, rather than before, installing Y. *two commas*
2	The program occasionally fails to **launch. This** may cause problems particularly when several other programs are open at the same time and thus memory is low.	The program occasionally fails to **launch, which** may cause problems particularly when several other programs are open at the same time and thus memory is low.
3	There are three advantages: • costs are lower • deadlines are more easily met • customers are generally happier	There are three advantages: costs are lower, deadlines are more easily met, and customers are generally happier.
4	There are three advantages of this: costs are lower, deadlines are more easily **met, and customers** are generally happier.	There are three advantages of this: costs are lower, deadlines are more easily **met and customers** are generally happier.

Commas should also be used:

1. to separate two dependent clauses. This is often the case with clauses introduced by *if, when, as soon as, after* etc. It also applies when the second clause is introduced by *then* or *which*
2. after sentences that begin with a link word that indicates you are adding further information or talking about a consequence. Examples: *also, moreover, in addition, furthermore, however, despite this, on the other hand, consequently*

16.3 Commas (,) (cont.)

3. with whole numbers. But with decimals use a point (.)
4. before *for example*. Don't have an example in the middle of a sentence, so begin a new sentence after you have given an example

	Yes	No
1	If the program fails to **launch,** call the help desk.	If the program fails to **launch** call the help desk.
1	The program occasionally fails to **launch,** which may cause problems.	The program occasionally fails to **launch** which may cause problems.
2	In addition, you can … **Consequently,** you can also …	In addition you can … **Consequently** you can also …
3	Our products are used by over **10,000** traders worldwide and represent **27.5%** of the total …	Our products are used by over **10.000** traders worldwide and represent **27,5%** of the total …
4	This application can be used on most platforms**, for** example XTC and B4ME. **It** can also be used with ..	This application can be used on most platforms **for** example XTC and B4ME, **it** can also be used with ..

Do not use a comma when you have a series of nouns where the first and second noun are not related. Instead, begin a new sentence after the first noun, otherwise the reader will think that the nouns are all part of the same series.

Yes	No
Each row in the page represents an **image. The** information and the features provided enable you to resize, crop and edit the images created.	Each row in the page represents an **image, the** information and the features provided enable you to resize, crop and edit the images created.

16.4 Hyphens (-)

Consider turning off the automatic hyphenation function. A text is much easier to read if lines do not end in hyphens.

Yes	No
These are inserted **automatically** at the end of lines.	These are inserted **auto-matically** at the end of lines.

Use a hyphen when you:

1. join two nouns together to form an adjective to describe another noun. Note: do not use a plural *s* on the noun that is acting as an adjective (i.e. the first noun)
2. use a word that acts as a prefix to the following word
3. you precede a word with *non*
4. join a noun to a preposition (*clean-up, back*-up), but do not join a verb to a preposition (*to clean up, to back up*). There are no clear rules regarding which nouns should be joined to their related preposition – if in doubt, check on Google!

See also 21.5.

	Yes	No
1	A 30-**year**-old manager.	A **30 years old** manager.
1	**Row-based** flashing instead of cell-based.	**Row based** flashing instead of cell based.
2	Control of the interaction is **user-not application**-driven.	Control of the interaction is **user not application** driven.
3	These are **non-essential** items.	These are **non essential** items.
		These are **nonessential** items.
4	When you **start up** the machine, make sure ...	When you **start-up** the machine, make sure ...
	This feature is only available at **start-up**.	

16.5 Parentheses ()

When readers see a phrase in parentheses, they may assume that the information contained therein is not very important. Don't use parentheses when it would be less distracting for the reader if you used a separate phrase.

Use parentheses:

1. with acronyms and abbreviations. Put the full form outside the parentheses, and the acronym inside
2. to give examples in the form of short lists, when this list appears in the middle of the phrase

Yes	No
This is based on a **first in first out (FIFO)** policy.	This is based on a **FIFO (first in first out)** policy.
This is only true of three **countries (i.e.** Libya, Syria and Jordon) **and** for the purposes of this operation it can be ignored.	This is only true of three **countries i.e.** Libya, Syria and Jordon **and** for the purposes of this operation it can be ignored.

16.6 Periods (.)

As a general rule:

1. do not use periods at the end of titles or headings
2. if a word like *etc* appears at the end of a sentence it only requires one period, never two
3. do not use a set of three (or more) periods to indicate *etc*

	Yes	No
1	A guide to the use of English in user manuals	A guide to the use of English in user manuals.
2	Various grammatical points are covered: tenses, adjectives, agreement etc.	Various grammatical points are covered: tenses, adjectives, agreement etc..
3	Various languages can be used, for example C++ and Java, and most types of hardware, for example IBM and Apple.	Various languages can be used (C++, Java, ...) on most types of hardware (IBM, Apple, ...).

16.7 Semicolons (;)

You should never need to use a semicolon in a manual:

1. do not join two independent clauses with a semicolon. Instead, make two simple, separate sentences
2. when several items are mentioned, use a list

	Yes	No
1	Users can now see the entire **database. There is also a** special alert mechanism to inform administrators …	Users can now visualize the entire **database;** **a** special alert mechanism is also provided that informs that administrator …
2	The following were installed: • X and Y • Q, R and S • B and C • D	We installed X and Y; Q, R and S; B and C; and D.

16.8 Forward slash (/)

Use a space before and after a forward slash. For example, *and / or* not and / or

17 CAPITALIZATION

Use initial capitalization to help the reader navigate the document by enabling them to see key words that might otherwise not be seen when skimming the document.

17.1 Titles of documents

Use initial capital letters (upper case letters) for all the words in the main title of a document. However, the words below should be in lower case, unless they are the first word:

- *a* and *the*
- *it*
- *and*
- all prepositions (*by, from, of* etc)

Do not use a full stop (.) at the end of a title.

Alternatively, only use initial capitals for the first word, and for any other words that would normally require initial capitalization.

Yes	No
A Guide to the Use of English in User Manuals	A Guide To The Use Of English In User Manuals.
A guide to the use of English in user manuals	

17.2 Section headings

Just use an initial capital letter for the first word. Then all the other words in lower case.

Do not use a full stop (.) at the end of a heading. In the example below, *installing the software* is the heading of a section.

Yes	No
Installing the software	Installing the Software.
The software requires version 2.1, or higher, of…	The software requires version 2.1, or higher, of…

17.3 Product names

Product names – both of your own company and of others – generally have initial capital letters. In any case make sure you reproduce them correctly.

17.4 Days, months, countries, nationalities, natural languages

Days, months, countries, nationalities and languages all have an initial capital letter.

Yes	No
The new version in **English** will be released on **Monday**, 10 **March** throughout **Europe** and **Asia**.	The new version in **english** will be released on **monday**, 10 **march** throughout **europe** and **asia**.

17.5 Notes

Use a capital letter after *Note*.

Yes	No
Note: The latest version of KwikTrans must be installed for this function to work optimally.	Note: the latest version of KwikTrans must be installed for this function to work optimally.

17.6 Acronyms

All the letters of acronyms have capital letters.

Yes	No
The application enables traders to trade on **LIFFE** and **NASDAQ**.	The application enables traders to trade on **Liffe** and **Nasdaq**.

17.7 OK

Both the letters in OK are capitalized.

Yes	No
Click **OK**.	Click **ok**./ Click **Ok**.

17.8 Figures, tables, sections

Figures, tables, sections and similar, have special rules for initial capitalization:

1. when you refer to numbered *sections, figures, tables, appendices, schedules, clauses* etc, capitalize the initial letter
2. do not capitalize the initial letter of *section, figure* etc when there is no number associated
3. *the* is not required when *sections, figures* etc have an associated number
4. the text is clearer for the reader if you do not use abbreviations such as *Fig.* and *Sect.*

	Yes	No
1	See **Section** 2 for further details.	See the **section** 2 for further details.
2	See the **appendix** for further details.	See the **Appendix** for further details.
3	The specific architecture is shown in **Fig. 5**.	The specific architecture is shown in **the Fig. 5**.
4	The specific architecture is shown in **Figure 5**.	The specific architecture is shown in **Fig. 5**.

17.9 Steps, phases, stages

Capitalize the initial letter of *step, phase* and *stage*, when there is a number associated.

Yes	No
See **Step** 1 above.	See **step** 1 above.
These tasks will be accomplished in **Phase** 2 of the project.	These tasks will be accomplished in **phase** 2 of the project.

17.10 Keywords

In some documents, such as specifications and contracts, you may need to distinguish between different users, projects, products and other things. In such cases, initial capitalization is useful to make these keywords stand out for the reader.

Yes	No
There are two types of user. Hereafter they will be referred to as **User A** and **User B**.	There are two types of user. Hereafter they will be referred to as **user a** and **user b**.
The two parties shall be referred to as the **Vendor** and the **Supplier**.	The two parties shall be referred to as the **vendor** and the **supplier**.
In the first phase, two products will be sold: a product for automatically connecting to banks (**Product** 1), and a product for risk management (**Product** 2).	In the first phase, two products will be sold: a product for automatically connecting to banks (**product** 1), and a product for risk management (**product** 2).

18 ABBREVIATIONS AND ACRONYMS

18.1 Limit usage of abbreviations

An abbreviation is the short form of word (*info* for *information*). Abbreviations are not usually used in documents because they are a sign of informality and they are often less readable. Note that words such as *figure, table, appendix*, are used with initial capitalization (i.e. *Figure* rather than *figure*) when they are associated with a number.

Yes	No
See Figure 5 on page 10.	See fig. 5 on p. 10.
See the figure below.	See the fig. below.

18.2 Quantities

Abbreviations of quantities (examples: *meters, kilograms*) are not followed by a period (.). The number that comes before them is generally followed by a space. Write such abbreviations in lower case.

Yes	No
The road is **3 km** long.	The road is **3 km**. long.
	The road is **3 KM** long.

18.3 Introducing an acronym

When introducing an acronym for the first time, use the full form and put the acronym in brackets. Afterwards, just use the acronym.

If you use a lot of uncommon acronyms, provide a glossary.

Yes	No
Orders are dealt with on a **first in first out (FIFO)** basis.	Orders are dealt with on a **FIFO (first in first out)** basis.

18.4 Punctuation

The number that comes after an abbreviation is preceded by a space.

When associated with a number, abbreviations generally have an initial upper case letter. An exception is page, which is always lower case (example: *the figure on p. 1*).

Do not insert full stops (.) into acronyms.

Yes	No
See Fig. 1 on p. 2.	See fig. 1 on P. 2
The program should be installed directly onto the user's PC.	The program should be installed directly onto the user's P.C.

Plurals

Here are some examples of common abbreviations in their singular and plural forms.

Singular	Plural
See Fig. 5.	See Figs. 1-5.
It is 1 km from here.	It is 20 km from here.
See the figures on p. 4.	See the figures on pp. 4-7.
Item No. 1.	Item Nos. 1 and 5.

Make acronyms plural by adding a lower case *s*. Do not use apostrophes.

Yes	No
There are three **PCs** available.	There are three **PC** available.
	… three **PC's** available.
	… three **PCS** available.

18.5 Duplication

Do not repeat the final abbreviated word in the text following the abbreviation. In the examples below, the I in GUI stands for *interface*, and the N in PIN for *number*.

Yes	No
The GUI is user friendly.	The GUI **interface** is user friendly.
Insert your PIN.	Insert your PIN **number**.

19 BULLETS

19.1 Types of bullets

Use the same type of bullets consistently. Use:

1. round bullets (or other non-numbered bullets, e.g. dashes [-]) when the sequence of the items is not important
2. numbered bullets when the sequence of the items is important
3. numbered bullets for procedures

	YES	NO
1	To install the system you need: • Version 5.6 or later of Technophobe • Version 1.2 or later of Monstermac • Version 9.7 or later of SysManiac	To install the system you need: 1. Version 5.6 or later of Technophobe 2. Version 1.2 or later of Monstermac 3. Version 9.7 or later of SysManiac
2	The project is organized into three phases: 1. Specifications 2. Design and development 3. Release	The project is organized into three phases: • Specifications • Design and development • Release
3	To replace a word or phrase: 1. Select Replace from the Edit menu 2. Type in the word you want to replace 3. Click OK	To replace a word or phrase: √ Select Replace from the Edit menu √ Type in the word you want to replace √ Click OK

Note

Tick (√) bullets are not normally used in manuals. Instead they are used to highlight what particular features a product has. They are also used to indicate what has been done to fulfill a request from a client. In any case they can be replaced with normal round bullets.

19.2 When to use

Use bullets rather than continuous text to help the reader understand a:

1. procedure to follow
2. list of requirements
3. list of events
4. list of people, places etc

	Yes	No
1	3.7 MANAGING MANUAL TRANSLATIONS To execute a new manual translation: 1. Open the Manual Trans dialog in the KT component. 2. Click the Manual Trans button on the KT Controller.	3.7 MANAGING MANUAL TRANSLATIONS To execute a new manual translation, use the Manual Trans dialog in the KT component. To open the Manual Trans dialog, click the Manual Trans button on the KT controller.
2	This component requires the following hardware: • XYZ version 7.1 or later • PQR • ABC	This component requires an XYZ (version 7.1 or later), a PQR and an ABC.
3	The following sequence of events occurs as a result of this virus: - Random characters begin to appear in the text. - The specific program blocks. - All programs block. - The computer crashes.	As a result of this virus, first random characters begin to appear in the text, then the specific program blocks, after that all programs block, and finally the computer crashes.
4	This guide is intended for: • professional translators • EFL teachers These include ESP, EAP, TESOL and business English. • simultaneous interpreters	This guide is intended for professional translators, EFL teachers (these include ESP, EAP, TESOL and business English, and simultaneous interpreters.

Note: As highlighted in the second bullet of the fourth example, bullets can be made up of more than one paragraph. This helps the reader to understand the content more quickly.

19.3 Punctuation

There are no standards for the use of punctuation in bullets. In this book I have used both of the two styles below.

Style 1

- use a colon (:) at the end of the introductory phrase to the bullets
- begin each bullet with a lower case letter
- do not use any punctuation at the end of each bullet, including the final bullet

Style 2

- Use a colon (:) at the end of the introductory phrase to the bullets (as in the first style);
- Begin each bullet with a lower case letter;
- End of each bullet with a semicolon (;), and put a period (.) at the end of the final bullet.

19.4 Introducing bullets

Note how *the following* and *as follows* are used to introduce bullets:

Yes	No
The following fields will be shown: - name - organization - address	They will be shown the following: - name - organization - address
The calculation method is as follows: 1. one 2. two 3. three	There follows the calculation method: 1. one 2. two 3. three

19.5 Avoid redundancy

Don't repeat words unnecessarily. Incorporate repeated words into the introductory sentence.

Yes	No
2.3 FEATURES KwikTrans will translate: • up to 100 pages in less than 10 s • into an unlimited number of languages	2.3 FEATURES KwikTrans has the following features: • **it will translate** up to 100 pages in less than 10 s • **it will translate** into an unlimited number of languages

19.6 Bullets after section titles

Experts recommend not using bullets immediately after a section title. Instead, bullets must be introduced by some text.

Yes	No
2.3 FEATURES KwikTrans has the following features: • Translates up to 100 pages in less than 10 s. • With one click, translates into an unlimited number of languages.	2.3 FEATURES • Translates up to 100 pages in less than 10 s. • With one click, translates into an unlimited number of languages.

However such introductory texts may be unnecessary and become cumbersome, so I have ignored this 'rule' myself in many parts of this book.

19.7 One idea per bullet

Each bullet should only contain one idea.

Yes	No
There are **four** ways to improve your English: - **Find a good teacher** - **Go to lessons** - Watch YouTube - Use Google Translate	There are **three** ways to improve your English: - **Find a good teacher. Go to lessons** - Watch YouTube - Use Google Translate

19.8 Grammatical consistency

Within the same list or set of bullets, always begin with the same grammatical form.

Use an introductory phrase that can always be followed by the same grammatical type (preferably an infinitive or a gerund).

Yes	No
P is used to: - acquire X - send Y - receive Z	P is used: - for the acquisition of X - to send Y - for receiving Z
P is used for: - acquiring X - sending Y - receiving Z	

20 FIGURES, TABLES AND CAPTIONS

20.1 Making reference to figures

Figures are common in manuals. Here are some guidelines:

1. use *figure* to refer any kind of picture, screenshot or diagram in your manual
2. always use the same verb - *show* - when referring to what is contained in a figure. It is NOT necessary to use a series of synonyms such as *outline, sketch, display, highlight, evidence*
3. be concise when introducing the figure
4. where possible use the active form rather than the passive (see Chapter 30)
5. use *as* not *as it* (see 32.6)
6. when the figure immediately follows the introductory text, then end the introductory text with a colon (:)

	Yes	No
1	The **figure** below shows the underlying architecture.	The **schema** below shows the underlying architecture.
2	The figure below **shows** the initial settings.	The figure **sketches** the initial settings.
3	**Figure 2** below **shows** the initial settings.	The following figure (Figure 2) **gives a schematic overview** of the initial settings.
4	Figure 2 below **shows** the initial settings.	The initial settings **are shown** in Figure 2 below. In Figure 2 the initial settings **are shown**.
5	**As** can be seen in the figure below:	**As it** can be seen in the figure below.
6	The figure below shows the quarterly yield**:**	The figure below shows the quarterly yield**.**

20.2 Numbering figures

If your document only has two or three figures in it, it is probably not worth numbering the figures. Simply refer the reader to the figure 'above' or 'below'.

If you decide to number the figures:

1. do not use *the* before *figure, appendix, table, schedule* when they are followed by a number
2. words such as *figure, appendix, table, schedule* tend to have initial capitalization when followed by a number (see 17.8). This helps to make them stand out for the reader
3. use *the* before *figure, appendix, table, schedule* when they are not followed by a number. In such cases, do not use an initial capital letter

	Yes	No
1	**Figure** 3 shows that 80% of users prefer this system.	**The figure** 3 shows that the 80% of users prefer this system.
2	See **Appendix** 2 for details.	See **appendix** 2 for details.
3	As can be seen in **the figure** below, the higher values…	As can be seen **in Figure** below, the higher values…

20.3 Abbreviations with figures, tables, appendices etc

Only use abbreviations for words such as figure, table, appendix, when such words are associated with a number.

Yes	No
See the figure below.	See the fig. below.
See Fig. 1 below.	See fig. 1 below.

20.4 Captions to figures

Use the following format:

> Figure 1. Network architecture.

That is to say:

- do not use an abbreviation
- initial capital letter for figure, table, appendix etc
- after the number put a full stop
- initial capital letter for the first word in the description
- end the line with a full stop

Yes	No
Figure 1. Network architecture.	Fig 1. Network architecture
	figure 1. network architecture
	Figure 1 Network Architecture

20.5 Use tables to show information quickly and clearly

Tables should be easy to understand. Below is an example from a software manual:

This event	Means this
Update	A record update was received from KwikTrans.
Sending	KwikTrans is sending the record data to the queues.
Error	KT MQ is signaling an error in the 'accents' operation.
OK	KT MQ is signaling a successful 'accents' operation.

21 DATES AND NUMBERS

21.1 Day / Month / Year

Write the day as a number, the month as a word, and the year as a number. This avoids problems with the US system of putting the month first (9/11 = the eleventh of September in the US).

Do not use *1st, 2nd, 3rd, 4th* etc. They add no useful information and you may inadvertently use the wrong form.

the and *of* are not required when writing the date, but only when speaking.

Yes	No
The product will be released on **10 March 2020**.	The product will be released on **March 10, 2020**.
	… on **10 / 3 / 2020**.
	… on **the 10th of** March 2020.

21.2 Decades

When referring to decades (periods of ten years), use the full numerical form.

Do not use apostrophes.

Yes	No
The company began selling this product in the **1990s**.	The company began selling this product in the **'90s**.
	… in the **1990's**
	… in the **nineties years**.

21.3 Words (*twelve*) vs digits (12)

Although there are no strict rules, here are some guidelines for deciding whether to use a word or a digit:

1. do not begin a sentence with a number written as numerals
2. if possible, rearrange the sentence so that the number does not appear at the beginning
3. if it is not possible to apply Rule 2, use words instead
4. do not mix words and figures to refer to the same number
5. when you use numbers from one to ten within a written text, write them as words (e.g. *nine*) rather than figures. (e.g. 9)

	Yes	No
1	**Two hundred** new releases will be made in April.	**200** new releases will be made in April.
2	This feature is not used by **50%** of users.	**50%** of users do not use this feature.
3	**Fifty per cent** of managers believe that…	**50%** of managers believe that…
4	There were **200,000** people at the conference. There were **two hundred thousand** people at the conference.	There were **200 thousand** people at the conference.
5	There will be **nine** new releases in April.	There will be **9** new releases in April.

Note:

Exceptions to Rule 5:

- A series of numbers. Example: *In the last three years the numbers have risen by 11, 6 and 7, respectively.*
- When numbers act as adjectives. Examples: *a 4-point plug, a size 5 component, a 7-year-old child*

21.4 Points vs commas

Write decimals using a point (.).

Write whole numbers using a comma (,).

Yes	No
There are **20,000** clients using this product, representing **56.54%** of the market.	There are **20.000** clients using this product, representing **56,54%** of the market.

21.5 Ranges, fractions, periods of time

To indicate a range of values with:

1. figures use a hyphen (-)
2. words, use *to*

Use a hyphen (see 16.4) with:

3. fractions that are made up of two words (e.g. *three-fifths, seven-ninths*)
4. ages and periods of time. Note that there is no plural *-s* in *day, week, year* etc when these words act as adjectives

	Yes	No
1	The course will last **15–20** weeks.	The course will last **fifteen – twenty** weeks.
2	The course will last **three to four** weeks.	The course will last **three – four** weeks.
3	**Three-quarters** of the employees in this company come to work by car.	**Three quarters** of the employees in this company come to work by car.
3	**Four-week** holidays can only be taken by **40-year-old** employees.	Four **weeks** holidays can only be taken by 40 **years** old employees.

Note:

You can introduce a range of values in three different ways:

*There should be **11–20** participants.*

*There should be **from 11 to 20** participants.*

*There should be **between 11 and 20** participants.*

21.6 Percentages

Be careful when using percentages:

1. do not begin a sentence with a percentage expressed in numbers. Use words instead
2. do not put *the* before percentages
3. *Percentage* is one word. Do not write *%age* or *per centage*
4. to save space, where possible (see Rule 1) write percentages using numbers (70%) rather than words (seventy per cent)

	Yes	No
1	Eighty **per cent** of users prefer this system.	**The** 80% of users prefer this system. **The** eighty **percent** of users prefer this system.
2	Figure 3 shows that **80%** of users prefer this system.	Figure 3 shows that **the** 80% of users prefer this system.
3	The value is expressed as a **percentage**.	The value is expressed in **percentage**. The value is expressed as a **%age**.
4	Figure 3 shows that **80%** of users prefer this system.	Figure 3 shows that **eighty per cent** of users prefer this system.

22 GIVING EXAMPLES

22.1 *for example*

Below are some rules for the use of *for example*:

1. use a comma before *for example*
2. if you have a list (two or more) of examples, then begin a new sentence after these examples
3. don't use both *such as* and *for example* together. Use one or the other
4. *for instance* and *like* are not normally used in technical documents. Use *for example*
5. if you write *for example* after the example, rather than before, then it should be preceded and followed by commas. *for example* should never be placed at the end of the phrase

	Yes	No
1	Whenever you use your **PIN, for** example to get money from an ATM, do not let anyone see you.	Whenever you use your **PIN for** example to get money from an ATM, do not let anyone see you.
2	This application can be used on most platform**s, for** example XTC and B4ME. **It** can also be used with…	This application can be used on most platform**s for** example XTC and B4ME, **it** can also be used with…
3	Our company has offices in many countries in Europe, **for example** France and Spain.	Our company has offices in many countries in Europe, **such as for example** France and Spain.
4	Our company has offices in many countries in Europe, **for example** France and Spain.	Our company has offices in many countries in Europe, **like / for instance** France and Spain.
5	Many governments are in crisis. In **Italy, for example, the** government is facing…	Many governments are in crisis. In **Italy for example the** government is facing. Many governments are in crisis. In Italy the government is facing big problems with the unions,**for example.**

22.2 *e.g., i.e.*

Consider not using *e.g.* and *i.e.* (or *eg* and *ie*) in technical documents because they are often confused. Instead use *for example* and *that is to say*, respectively

If you do use *e.g.* and *i.e.* in other documents, for example emails, then make sure you know the difference between them:

e.g. introduces an example: *Many countries in Europe, e.g. Spain, France and Germany.*

i.e. specifies: *The UK is made up of four countries,* **i.e.** *England, Scotland, Wales and N. Ireland.*

22.3 *etc*

When you introduce a series of examples:

1. if possible, think of some more meaningful than *etc*
2. with *for example*, do not put *etc* at the end

	Yes	No
1	This is true in many nations, for example Spain, Germany, Italy, the UK **and other European countries**.	This is true in many nations, for example Spain, Germany, Italy, the UK **etc**.
2	This is true in many countries, for example Spain, Japan and Togo.	This is true in many countries, for example Spain, Japan, Togo **etc**.

22.4 Dots (...)

Do not use a series of dots (...) at the end of a list of examples to indicate *etc*. Instead, begin the list with *for example*.

Yes	No
This is true in many nations, for example Spain, Germany, Italy, the UK and other European countries.	This is true in many nations (Spain, Germany, Italy, the UK,...).

23 REFERENCING

23.1 Sections and documents

When you refer to another section or document use this format:

1. use the least number of words when you make references to other sections within the same document or to external documents. Do not use words like *document, guide, book* unnecessarily

2. do not use *above* and *below* when you refer to something that appears later in the same document, unless you are talking about a figure or table that immediately follows. Likewise, rather than saying *on the previous page* or *in the next section*, help the reader by sending him / her to a specific page number or heading

3. do not simply put the section number. Put the section title as well. This will help the reader decide whether they will really need to go to that section for more information

Yes	No
For details, see page 12 of the KwikTrans User Guide.	**For further details, the reader can refer to** page 12 of the **document entitled** KwikTrans User Guide.
See 'Configuration' on **page 26**.	See 'Configuration' **below**.
See Section 2 *Installation*.	See Section 2.

23.2 Figures, tables, windows

Use the following introductory verbs:

> A figure **shows** something or **illustrates** something.
>
> A table **lists** values. A table **provides** information on something.
>
> A window is **displayed** and **shows** a value.

When you want to introduce a figure or table that immediately follows the text:

1. write: *the following figure* or the *figure below*
2. end the introductory sentence with a colon (:)
3. prefer an active to a passive form
4. do not use terms such as *schema*, *screen cap* or *screenshot*. Just use *figure* or *picture*
5. give the page number if the figure or table does not immediately follow the text but is on another page

	Yes	No
1	The **following** figure shows the new platform:	The new platform is **as follows**.
2	The figure below shows the new **platform:**	The figure below shows the new **platform.**
3	The figure below **shows** the new platform:	The new platform **is shown** in the figure below.
4	The architecture is simple, as highlighted by the **figure** below.	The architecture is simple as highlighted by the **screenshot** below.
5	The architecture is simple, as highlighted by the figure **on page 21**.	The architecture is simple, as highlighted by the figure **below**.

23.3 *the following*

Note that *the following* is followed by a noun (*versions* in the example below).

Yes	No
The following versions can be used:	The versions that can be used are the following:

23.4 *above mentioned / as mentioned above*

When you want to refer back to something you wrote about in the previous paragraph:

1. use: noun + *mentioned above* or *above-mentioned* + noun
2. however, it is much clearer simply to repeat the reference because then readers will not have to re-read sentences in order to understand.
3. the phrase *as mentioned above* is not normally helpful for the reader. Either eliminate it completely or replace it with a direct reference to the section and paragraph where the thing is mentioned

Yes	No
The **function mentioned above** should only be used when…	
The **above-mentioned function** should only be used when…	The **function above mentioned** should only be used when…
The **Buy/Sell function mentioned above** should only be used when…	The **above-mentioned function** should only be used when…
This procedure is…	**As mentioned above**, this procedure is…
As mentioned **in Section 2**, this procedure is…	As mentioned **above**, this procedure is…

Note:

The same rules apply to *below* and *see below*.

Yes	No
This procedure is described in **Section 4**.	This procedure is described **below**.
This procedure is extremely complex and **is described in Section 4**.	This procedure is extremely complex – **see below**.

23.5 *hereafter*

Hereafter is a useful word when you have some terminology that you want to abbreviate and then use the abbreviation in the rest of the document.

Yes	No
This feature is known as an 'automatic rendering and masking agent' **hereafter** ARM agent.	This feature is known as an 'automatic rendering and masking agent' **in the rest of the document it will be referred to as the** ARM agent.

24 SPELLING

24.1 US vs GB spelling

Most international companies use US spelling rather than GB spelling. So set your spell check to 'English US'. This will then help you to decide whether, for example, to write *realise* (GB) or *realize* (US). There are no accurate rules on when to spell a word –IZE rather than –ISE, but Word's US spell check prefers –IZE.

Make sure your spelling is consistently British or American. Always use the spelling utility on your computer to check your spelling, although remember it won't check everything (see 24.3).

Some words that are frequently found in manuals:

US spelling: *behavior, catalog, center, color, modeled, modeling, program, signaling, traveled, traveling*

GB spelling: *behaviour, catalogue, centre, colour, modelled, modelling, programme, signalling, travelled, travelling*

24.2 Technical words

Spell checkers only highlight words that are not listed in their dictionaries. This means that you will manually have to check the spelling of some technical words, names of products etc.

24.3 Misspellings that automatic spell checkers do not find

Some of your misspellings will not be highlighted because they are words that really exist. When you have finished your document, do a 'find' and check if you have made any of the mistakes listed below. The first word in each case is probably the word you wanted to use:

> addition vs addiction, assess vs asses, context vs contest, drawn vs drown, thanks vs tanks, though vs tough, through vs trough, two vs tow, use vs sue, which vs witch, with vs whit

Other typical typos include:

> chose vs choice, form vs from, filed vs field, found vs founded, lose vs loose, relay vs rely, than vs then, three vs tree, weighed vs weighted

Part IV
Typical Mistakes

The following chapters outline some mistakes (in the NO columns in the tables) in terms of vocabulary and grammar. The mistakes mentioned are by no means exhaustive.

25 COMPARISONS

25.1 Comparative vs superlative

1. use the comparative form (*more, better, worse*) to compare two things
2. use the superlative form (*most, best, worst*) to describe something in absolute terms

Yes	No
1 The system performed **better / worse / more efficiently** in the first test than in the second test.	The system performed **best / worst / most efficiently** in the first test than in the second test
2 The application returns only the **most** relevant results.	The application returns only the **more** relevant results.
2 It always chooses the **best** solution.	It always chooses the **better** solution.

25.2 Adverbs and prepositions used with comparisons

Note the use of:

1. *than* not *then* or *of* when comparing two things or people
2. *as* not *of* with *the same*
3. *as... as* in comparatives that highlight the similarity between things (in both affirmative and negative sentences)

Yes	No
1 The first is better **than** the second.	The first is better **then / of** the second.
2 The first is **the same as** the second.	The first is **the same of** the second.
3 The first is **as good as** the second.	The first is **good as** the second.
3 The first is **not as good as** the second.	The first is **not so good like** the second.

When you make a comparison between two or more things use *than* rather than *with respect to / in comparison to / compared to*. *Than* is much shorter.

Yes	No
X is bigger **than** Y.	X is big **with respect to** Y.

25.3 *the more... the more*

1. Note that, as in English in general, the verb in a comparison appears after the subject and not before
2. The definite article is required before each comparative
3. On some occasions, no verb is required

Yes	No
1 In realistic conditions, the more **robust the software is** the fewer problems there are.	In realistic conditions, the more **is robust the software**, the fewer problems there are.
2 **The more** you use the software, **the easier** it becomes.	**More** you use the software, **easier** it becomes.
3 The sooner the job is done, **the better.**	The sooner the job is done, **better is.**

26 DEFINITE ARTICLE (*THE*), INDEFINITE ARTICLE (*A, AN*), *ONE*

26.1 Definite article: main uses

1. DO NOT use **the** if you are talking about something in general and the noun is:

   in the plural (e.g. *computers, books*)

   uncountable (e.g. *hardware, information*) – see 26.4.

2. Use **the** to talk about something specific.

3. In phrases with the following combination: **singular noun + of + noun**, the first noun is nearly always preceded by **the**.

	Yes	No
1	Google do not sell **computers**, they sell **advertising**.	Google do not sell **the computers**, they sell **the advertising**.
2	**The computers** in our company are all Hewlett Packard, and **the software** used (by our company) is all proprietary software.	**Computers** in our company are all Hewlett Packard, and **software** used (by our company) is all proprietary software.
2	**All the computers** that we use are Hewlett Packard.	**All computers** that we use are Hewlett Packard.
3	**The aim** of this document is…	**Aim** of this document is…

26.2 Definite article: other uses

Do not use *the* before:

1. percentages

2. the following words (and similar words) when they are followed by a number:

    figure, appendix, table, schedule

    step, phase, stage

    question, issue, task

    case, example, sample

3. subjects of study (e.g. *computer sciences, engineering, economics*)

	Yes	No
1	Figure 3 shows that **80%** of users prefer this system.	Figure 3 shows that **the** 80% of users prefer this system.
2	**Figure 3** shows that 80% of users prefer this system.	**The Figure 3** shows that 80% of users prefer this system.
2	Details can be found in **Schedule 2.**	Details can be found in **the schedule** 2.
3	All the founders have degrees in **computer sciences.**	All the founders have degrees in **the computer sciences.**

26.3 *a / an* + **singular countable nouns**

A countable noun is something you can count: *30 books, many employees, 100 apples, several PCs.*

Before a singular countable noun you must put an article (*a / an* or *the*).

Yes	No
As **a** rule, the system works best if..	As rule, the system works best if…
It acts like **a** checker…	It acts like checker…

Note:

Do not use *a / an* with *input, output,* and *license / licence.*

We used X as input, and Y as output.

Under license from ABC

With *probability* the indefinite article is optional:

... with probability 0.25

... with a probability of 0.25

26.4 Uncountable nouns

An uncountable noun is seen as a mass rather than as several clearly identifiable parts.

You cannot use *a / an* or *one* before an uncountable noun.

An uncountable noun cannot be made plural – use a singular verb and do not put an *s* at the end.

You cannot say:

> all the / three / several / many / informations
>
> an / another / one / each / every information
>
> an English / one English / three Englishes

But you can say:
> a lot of information, some information, not much information

Examples of uncountable nouns frequently used in technical manuals, brochures, websites and newsletters:

> access, accommodation, advertising, advice*, agriculture (and other subjects of study), capital, cancer (and other diseases and illnesses), consent, electricity (and other intangibles), English (and other languages), equipment*, evidence*, expertise, feedback, functionality, furniture*, gold* (and other metals), hardware, health, industry, inflation, information*, intelligence, luck, knowhow, luggage*, machinery*, money, news, oxygen (and other gases), personnel, poverty, progress, research, safety, security, software, staff, storage, traffic, training, transport, waste, wealth, welfare, wildlife.

The uncountable nouns listed above with an asterisk (*) can be used with *a piece of*. This means that they can be used with *a / an, one* and be made plural. Examples: *a piece of advice, two pieces of equipment, one piece of information.*

Yes	No
The document gives **feedback / information** on the **performance** of the system.	The document gives **feedbacks / informations** on the **performances** of the system. The rest of the document gives **a feedback / an information** on…
This is an innovative **application**.	This is **an innovative software**.
This may cause a lot of **damage**.	This may cause a lot of **damages**.

26.5 *a* vs *an*

an is used before:

- *a, e* (but not *eu*) *i*, and *o*.
- *u* when the sound is like the *u* in *understanding, unpredictable*
- letters in acronyms which begin with a vowel sound
- before *hour* (and *herb, heir, honest, honor*) It is not used before other words that begin with H, unless the H appears in an acronym.

Examples:

a Sony laptop	**an** Apple laptop
a university, **a** USB	**an** understanding
a u-r-l (letters pronounced separately)	**an** url (letters pronounced as one word)
a European project	**an** EU project
a hierarchy	**an** hour
a Hewlett Packard computer	**an** HP computer

26.6 *a / an* vs *one*

one is a number (one, two, three). Use *one* instead of *a / an* when it is important to specify the number. Examples:

We need **one** manual not two manuals.

There is only **one** thing to do. [not two or three]

There is at least **one** other company in this field. [so there may be two other companies]

This parameter has **a** unique value. [something cannot have two unique values]

Use *one* instead of *a / an* in the following types of expression:

One way to do this...

From **one** place / problem etc to another...

But when *way* is preceded by an adjective use *a*

A good way to do this.

27 GENITIVE

27.1 General usage

The use of the genitive is very complicated and there are no clear rules. If in doubt, use Google to check the specific phrase that you want to use.

Below are some examples explaining where the genitive should and should not be used.

the customer's requirements is correct if it means one particular customer for whom your company is working at the moment. Here the definite article is used because we are referring to one specific customer.

our customers' requirements is correct because it refers to a specific set of customers who all belong to your company

customer requirements is correct if you are referring to all of your company's customers seen as a whole. In fact, in this case no definite article is required because we are referring to all customers, not specific ones.

customers requirements is wrong

27.1 General usage (cont.)

Generally speaking:

1. do not use the genitive with inanimate objects
2. only use the genitive with people, companies, and nations. If not use a noun + *of* + noun construction
3. however, when the person is a generic person and is not one particular person who we have in mind, then do not use the genitive

	Yes	No
1	The **user's** PC.	The **PC's** screen.
2	The screen of the PC.	The **PC's / PC** screen.
2	The pilot of the plane.	The **plane's** pilot
		The **plane** pilot.
2	The rules of mathematics.	Mathematics' rules.
3	We need to take into account **user needs**. i.e. all our generic users.	We need to take into account **user's needs**.
		We need to take into account **users needs**.
		We need to take into account **users' needs**.

27.2 Companies

Imagine your company is called ABC. Below are some examples of how you should use the genitive in association with your company's name.

ABC's founders / chairman / managers / employees

ABC's offices in Paris and Madrid

ABC's London office / ABC's office in London

ABC's products / services

In the first example above you could also say:

The founders / chairman / managers / employees of ABC.

Yes	Unusual or wrong
ABC's offices in Paris and Madrid.	ABC offices in Paris and Madrid.
ABC's offer.	The offer of ABC.
ABC products / ABC's products are renowned for their reliability.	The products of ABC are renowned for their reliability.
ABC's offer with regard to automotive products …	ABC offer with regard to automotive products …

Compare these two examples, which highlight the difference between the use and non use of the genitive in relation to a company name.

Benetton's decision to raise prices. [*the company*]

I never buy Benetton clothes. [*the type of clothes*]

27.3 Countries and towns

1. Use the genitive with countries
2. You cannot always use the genitive with towns. If you are not sure, use *the X of Y* construction

Yes	No
1	**Russia's** gold reserves.	The gold reserves **of Russia**. [possible but not common]
2	He studied computer science at the **University of Pisa.**	He studied computer science at **Pisa's University**.
	He studied computer science at **Pisa University**	

27.4 Periods of time

1. The genitive can be used with time periods.
2. But not when these are preceded by *a / the*.

Yes	No
1	I'm taking three **weeks'** vacation next month. = three **weeks** of vacation	I'm taking a vacation of three weeks next month.
2	He's on a 3-**week** vacation	He's on a three weeks' vacation.
	He's on a three-**week** vacation	

28 INFINITIVE VS GERUND

28.1 Infinitive

1. Use the infinitive when you talk about the aim / purpose of an action, or how to carry something out
2. Do not precede the infinitive with *for*
3. The negative infinitive is: *in order not to, so as not to*

Yes	No
1 **To make** money, our company designs and develops software.	**For to make** money, our company designs and develops software.
	For making money, our company designs and develops software.
2 I need money **to buy** a house	I need money **for buying** a house
3 **In order not to lose** data, make back-ups regularly.	**For not losing don't / To don't lose** data, make back-ups regularly.

28.2 Gerund vs infinitive

Note the difference between the gerund (activity) and the infinitive (purpose) in these examples:

Developing software is our primary activity.

Developing software requires specialized expertise.

To develop software you need specialized expertise.

(In order) to develop software, we employ top graduates.

28.3 Gerund

Use the gerund:

1. when the verb is the subject of the sentence
2. after a preposition
3. after *to* in the following cases: *to be dedicated to, to be an aid to, to look forward to, to be devoted to*

	Yes	No
1	**Developing** software is our core business.	**To develop** software is our core business.
2	Before **starting** up the PC, make sure it is plugged in.	Before **to start** up the PC, make sure it is plugged in.
3	This application is dedicated **to helping** traders.	This application is dedicated **to help** traders.

Do not use the gerund when:

1. you are simply giving additional information – in this case use *and*
2. you could replace the gerund with *that*
3. the clause introduced by the gerund explains the reason for the affirmation made in the previous clause – in this case use *since* or *because*
4. it is not clear what the subject of the gerund is. It is much clearer to use a subject + verb construction

	Yes	No
1	This document gives an overview of X **and throws** light on particular aspects.	This document gives an overview **of X, throwing** light on particular aspects.
2	Adrian only teaches students **that have** a good level of English.	Adrian only teaches students **having** a good level of English.
3	Adrian teaches students **because** he has a passion for teaching.	Adrian teaches students **having** a passion for teaching.
4	Once you'**ve started** the system, the bond … Once the system **has been started**, the bond …	After **starting** the system, the bond can be assigned to.

28.3 Gerund (cont.)

Note:

After *used*, you can put either the infinitive or *for* + ing

It is used **to write** code with.

It is used **for writing** code with.

28.4 *by* vs *thus* + gerund

1. *by* – to indicate how to achieve something, in this case the gerund is not the subject of the sentence (in the example below *you* is the subject):

    By clicking on the mouse you can open the window

    = **If** you click.

2. *thus* – to indicate the consequence of doing something:

    We learn English **thus** enabling us to communicate with our clients.

    = We learn English and **thus** we can communicate.

    = We learn English and **this means** we can communicate.

	Yes	No
1	**By learning** English you will pass the exam.	**Learning** English you will pass the exam.
2	We sell a lot of software **thus enabling** the company to pay its employees good salaries.	We sell a lot of software **enabling** the company to pay its employees good salaries.

29 NEGATIONS

29.1 Position

Negations (*no, don't, can't* etc) contain key information, so they should be located as near as possible to the beginning of the sentence.

1. If possible put the negation before the main verb. This immediately tells the reader that you are about to say something negative rather than something affirmative. It can be misleading to put the negation at the end
2. English tends to express negative ideas with a negation. This helps the reader to understand immediately that something negative is being said

	Yes	No, or not optimal *
1	We did **not** find **anything** to contradict these results.	We found to contradict these results **nothing**.
	We found **nothing** to contradict these results.	
1	Our results revealed that there is **no** relationship between X and Y.	Our results revealed that a relationship between X and Y does **not** exist. *
2	Two weeks is **not enough**.	Two weeks are **too few**.
2	We **don't** have **much** time available.	We have **little** time available. *

29.2 Contractions

Avoid contractions. Examples of contractions are: *don't, can't, won't*.

Negative information is generally very important: the negation stands out more clearly in the text if is written as separate words. If the information is absolutely vital, you can write NOT in capital letters.

Yes	OK, but may not be very clear
The system will not work when overloaded.	The system won't work when overloaded.
Do NOT turn your computer off.	Don't turn your computer off.

29.3 *no one* vs *anyone*

1. If you use *anyone* in a sentence that contains no negation, then it has a very similar meaning to *everyone* and thus has an affirmative meaning. This is true for all derivatives of *any*, e.g. *anything, anywhere*.
2. If the sentence has a negative meaning, use *not ... anyone* or *no one*
3. *without* and *hardly* do not require a negation

	Yes	No
1	**Anyone** can tell you that one plus one equals two.	
2	To the best of our knowledge **no one** has found similar results to these.	To the best of our knowledge **anyone** has found similar results to these.
	To the best of our knowledge there is**n't anyone** who has found ...	
3	You can do this **without any** problems or at least with **hardly any** problems.	You can do this **without no** problems or at least with **hardly no** problems.

29.4 Double negatives

1. Do not use *no* or *not* with negative words such as *no one, nowhere, nothing, neither, nor* etc
2. There is an exception to Rule 1: *not ... neither*. Note the word order and use of an auxiliary
3. It is easy to make mistakes with the construction explained in Rule 2, so it is better to use the normal construction

Note: *neither* and *nor* have the same meaning and can be used indifferently.

	Yes	No
1	We did **not** find **anything** to contradict these results, **either** in the first test **or** in the second test.	We did **not** find **nothing** to contradict these results, **neither** in the first test **nor** in the second test.
2	X did **not** function and **neither did Y.**	X did **not** function and **neither Y.**
3	X did **not** function and Y did **not** function **either**.	X did **not** function and Y did **not** function **neither**.

29.5 *also, both*

1. *also* and *both* are not generally found in negative sentences. Instead use *neither*.
2. You can use *both* with a negative only for contrast

Yes	No
1 **Neither** of them **functioned** as required.	**Both** of them **did not function** as required.
1 X did **not** function and **nor / neither** did Y.	X did **not** function and **also** Y.
2 We did **not** use **both** of them, just **one** of them.	

30 PASSIVE VS ACTIVE

30.1 Addressing the user

If your user guide was written specifically for only one type of user, then use:

1. *you* and active form to tell users what they can do (see 8.1 and 8.2)
2. the imperative to tell users how to do something

Readers will feel they are being addressed directly. Also, the phrases will generally be shorter.

	Yes	No
1	**You can create** new files.	New files **can be created**.
1	**You can customize** the user interface.	The user interface **can be customized**.
2	**Download** the key from our website.	The key **can be downloaded** from our website.

30.2 Referring to lists, figures, tables and documents

When you refer to lists, figures, tables etc (see Chapter 20), avoid using the passive. Use either:

1. *you*
2. an imperative
3. an active verb

	Yes	No
1	**You can find** more examples in Appendix 1	More examples **can be found** in Appendix 1.
2	**See** Section 3 for more examples.	More examples **can be found in** Section 3.
3	Table 1 **lists** all the values.	In Table 1 all the values **are listed**.
		In Table 1 **are listed** all the values.
3	The gateway **supports**:	The following features **are supported** by the gateway:

30.3 When the passive must be used

In English the verbs *install, upload* and *download* are not reflexive, and are generally used with the passive.

Yes	No
The system **is installed** automatically.	The system **installs** automatically.
Files **are downloaded** directly from source.	Files **download** directly from source.

Use the passive when you refer to things that your company has done and in any cases where it would sound strange to use *we*.

Yes	No
The software **was designed** for ease of use.	**We designed** the software for ease of use.

31 PRONOUNS

31.1 *you*

If you use *you*, the guide will become more user-friendly (see 8.1, 8.2 and 30.1). Readers will feel they are being addressed directly. Also, the phrases will generally be shorter.

Use active verbs (30.1, 30.2) with *you* as the subject, rather than impersonal passive verbs.

Yes	No
You can create new files.	New files **can be created**.
You can customize the user interface.	The user interface **can be customized**.

31.2 *we, us, our*

Do not use *we, us* or *our* in a user guide as a generic impersonal pronoun. Instead, use a phrase that does not require a pronoun.

Yes	No
This document **outlines** the main points of xyz.	In this document **we outline** the main points of xyz.
This means that **there are** two ways to do X.	This means that **we have** two ways to do X.
X functions as follows. First, it ...	**Let us now look** at the functioning of X. First, it ..

Use *you* rather than *we* or *us* if you are referring directly to the reader.

Yes	No
With this method **you** can do X.	With this method **we** can do X. This method allows **us** to do X.

31.3 he, she, they

1. Do not use *he* or *she* (*him, his, her, hers* etc) in a user guide. There is only one type of user and you should address that user using *you*
2. In other documents, avoid *he* and *she* (*him, his, her, hers* etc) by making the subject plural and using *they / their / them / theirs*
3. When the subject has to be singular, do not use just the masculine pronoun, instead use *his/her* and *he/she*

Yes	No
If **traders** are dealing on several markets and **they** wish to change **their** settings …	If **a trader** is dealing on several markets and **he** wishes to change **his** settings …
When **users** have some connection problems, the gateway tries to reconnect **them** automatically.	When **a user** has some connection problems, the gateway tries to reconnect **him** automatically.
There are two traders: Trader A and Trader B. If **Trader A** wants to send **his/her** order to the market then **he/she** has to …	There are two traders: Trader A and Trader B. If **Trader A** wants to send **his** order to the market then **he** has to …

31.4 *users*

1. Only use the term *user* when a document is intended for more than one type of reader.
2. In user guides, the word *user/users* should generally be replaced by *you*, because the reader is the user.

	Yes	No
1	Users should consult page 23 and systems administrators page 24.	
2	If **you** are making several purchases and **you** wish to change **your** settings for the addresses where the merchandise should be sent ...	If **the user** intends to make several purchases and **he / she** wishes to change **his / her** settings for the addresses where the merchandise should be sent ...
3	When **you** have some connection problems, the system tries to reconnect **you** automatically.	When **a user** has some connection problems, the system tries to reconnect **him / her** automatically.

When you are not writing a user guide, and you need to use terms such as *user, operator, technician*, try not to repeat them excessively.

Yes	No
Users can choose which control windows to show prices. There are two different layouts for this window. Fonts, features and settings can easily be configured.	**Users** can choose which control windows to show prices. **Users** can have this window in two different layouts. .... **Users** can easily configure fonts, features and settings...

31.5 *it, this*

1. Don't use *it* as the first word in a new section.
2. *this* must be accompanied by a noun.

In the examples below, INSTALLATION and NEW PRICE are headings.

Yes	No
1 INSTALLATION **The system** can be installed automatically.	INSTALLATION **It / This** is done automatically.
1 NEW PRICE **The New Price** configuration allows you to ...	NEW PRICE **It** allows you to ...
2 NEW PRICE **This configuration** allows you to ...	NEW PRICE **This** allows you to ...

31.6 *one, ones*

1. Do not use *one* as a generic pronoun.
2. Avoid using *one* and *ones* to refer back to a previous noun.

Yes	No
1 This feature would be useful in many cases.	**One** can think of many examples where this feature would be useful.
2 This is used for both **large and small sizes**.	This is used for both **large sizes and small ones**.
2 If you make a **purchase order**, and then you create a new order based on the **original purchase order**, then ...	If you make a purchase order, and then you create a new order based on the **original one**, then ...

31.7 *that, which, who*

that, which – for things

who – for people

If you put an adjective after the noun it describes, then this adjective should be introduced by *that/which/who*.

Yes	No
1	The company's CEO is Jan Garbarek, **who** is also one of the founders.	The company's CEO is Jan Garbarek **that** is also one of the founders.
2	I have several mobile phones, many **of which** don't work.	I have several mobile phones, many **of the which** don't work.
2	The company employees many people, most **of whom** are Polish.	The company employees many people, **of which** most are Polish.
3	We employed a secretary **who is** 25 years old. She wrote a document **which/that is** five pages long.	We employed a **secretary 25** years old. She wrote a **document five** pages long.

31.8 *that vs which*

that – when you want to define the preceding noun to differentiate it from another noun

which – to add additional information about the preceding noun. In this case you are not differentiating the noun but simply giving further details. *which* is generally preceded with a comma (,).

Example:

> In a horse race, the horse **that** wins the race gets the prize.
>
> That horse, **which** *I've owned for several years,* has never won a race.
>
> That horse usually wins, **which** *is why I have bet on it.*

In the first example, *that wins the race* is essential to understanding the sentence, it defines which horse I am talking about – you cannot logically say *The horse gets the prize*. In such cases, you should use **that** (though in non technical English, **which** is often used)

In the second and third examples, the parts in italics are not essential to understanding the sentence. They could be omitted – e.g. you can say *That horse has never won a race*. In this case, commas must be used, and only *which* can be used, not *that*.

If commas are needed and parenthetical / non-essential information is given, then *which* is used.

Yes	No
Our company has many offices. I work for the office **that** is in Prague.	Our company has many offices. I work for the office **which** is in Prague.
Our company, **which** is a software house, employees 100 people in Prague.	Our company, **that** is a software house, employees 100 people in Prague.
Our company sells a lot of software, **which** is how the company gets its money.	Our company sells a lot of software, **that** is how the company gets its money.

32 VOCABULARY

This chapter outlines some typical verbs, nouns and adverbs used in technical manuals.

For a comprehensive list of link words (e.g. *also, moreover, in addition, furthermore*) and the differences between them, see Chapter 13 in *English for Research: Usage, Style, and Grammar*.

32.1 *allow, enable, permit, let*

Meaning:

allow = the most commonly used in manuals

enable = allow, make operational, activate

let = allow, but tends to be used in informal documents

permit = used in more official language (e.g. what is permitted by the authorities)

Usage:

allow (and *permit* and *enable*) is used in three different ways:

1. X allows something / someone to do Y.
2. X allows Y to be done.
3. X is allowed to do Y.

Yes	No
1	This software **allows one / you / users to** carry out tasks more quickly.	This software **allows to** carry out tasks more quickly.
2	This software **allows tasks to be carried out** more quickly (by users).	This software **allows to carry out tasks** more quickly by users.
3	With this password **users are enabled to use** the system.	With this password **users enable to use** the system.

32.1 *allow, enable, permit, let* (cont.)

Notes:

let has a different construction, there is no *to* before the infinitive:

X lets something / someone do Y.

Sometimes you may wish to use an alternative construction for the sake of variety.

Yes	Alternative
This device **allows you to speak** in any language.	With / Using this device **you can speak** in any language. With / Using this device, any language **can be spoken**. **It is possible to speak** in any language using this device.
All the profiles are listed in the Profile Definitions page, which **allows you** to create and manage these profiles.	All the profiles are listed in the Profile Definitions page where **you can create** and manage these profiles.
The pages available in the Configuration folder **enable you** to set the automatic system behaviors and interface.	**Use** the pages in the Configuration folder to set the automatic system behaviors and interface.

32.2 *function, functionality, feature*

Meaning:

function: a named procedure that performs a distinct service.

functionality: an uncountable noun like *feedback* and *equipment*. You cannot say *a functionality* or *functionalities*. It means a set of features seen as whole. However, *functionality* is now often used like a countable noun.

feature: a feature is a characteristic of an application.

Usage:

The difference between *functionality* and *feature* is like the difference between *furniture* and *chair*. *Furniture* is the equivalent to *functionality* and means a set of chairs, tables, desks etc seen as one item. On the other hand, a chair or a desk is grammatically like a *feature*.

Yes	No
We are hoping to add more **functionality** in the next release.	We are hoping to add more **functionalities** in the next release.
We are going to add **three new features** in the next release.	We are going to add **three new functionalities** in the next release.

32.3 *(the) last, (the) next*

last: the time before the current one (e.g. the week before the current week)

next: the time after current one (e.g. the year after the current year)

the last / the next: a specific time in the past (see the penultimate example in the table below) or the future

When you use *last* and *next* in conjunction with *weeks, months, years* (i.e. the plural form used to indicate a period), you must insert *few* before the time period.

Yes	No
We will be releasing the document **next week**, on Monday in fact.	We will be releasing the document **the next week**, on Monday in fact.
We opened our offices in Russia **last year**.	We opened our offices in Russia **the last year**.
The last year that the Olympics were held in London was in **2012**.	**Last year** that the Olympics were held in London was in 2012.
The **last few years** have seen a considerable rise in the number of …	The **last years** have seen a considerable rise in the number of …

32.4 *login* vs *log in*, *startup* vs *start up*, etc

log in and *log on* (and similar terms) are always two words when they are verbs. When they are nouns or adjectives they can either be one or two words (there don't seem to be any standards for this).

Join a noun to a preposition (*clean-up*) but do not to join a verb to a preposition (*to clean up*). There are no clear rules regarding which nouns should be joined to their related preposition – if in doubt, check with Google!

Yes	No
First **log in** to your computer.	First **login** to your computer.
This can be done at **log in time / login time**.	
When you **start up** the machine, make sure …	When you **start-up** the machine, make sure …
This feature is only available at **start-up**.	

32.5 *and*

Although *and* is a very common word, it can often cause ambiguity, so consider following these guidelines:

1. only use *and* to link together two closely related ideas. If they are not closely related, use bullets or start a new sentence
2. to highlight the divisions between two groups of entities, use *along with*
3. in lists, put a comma before *and*
4. with long lists, use semicolons to show groupings. However, a better solution is often to use bullets

	Yes	No
1	This section describes how to: - install the software - ensure all data is protected - download security updates	This section describes how to install the software **and** the various security measures that should be taken to ensure all data is protected, along with how security updates can be downloaded.
1	This section describes how to install the **software. It** also details what security measures to take to ensure all data is protected, and how security updates can be downloaded.	
2	A and **B, along with C** and D, are the most used solutions.	A and **B and C** and D are the most used solutions.
3	There are three advantages of this: costs are lower, deadlines are more easily **met, and** customers are generally happier.	There are three advantages of this: costs are lower, deadlines are more easily **met and** customers are generally happier.
4	To do this you need the following items: a hammer and a nail; some wood and some glue; and time, luck, and a lot of patience.	To do this you need the following items: a hammer and a nail, some wood and some glue and time, luck and a lot of patience.
4	To do this you need the following items: - a hammer and a nail - some wood and some glue - time, luck and a lot of patience	

32.6 *as, as it*

Not the difference between *as* and *as it*:

1. in phrases like *as highlighted, as can be seen, as mentioned below*, the subject of the verb is implicit and is generally *we* or *you* (*as we have highlighted, as you can see, as we have mentioned*). In such cases *as* is not followed by *it*
2. if you put *it* after *as*, the meaning is *since / because*

	Yes	No
1	This is not true, **as is** evident from the figure.	This is not true, **as it is** evident from the figure.
1	**As can** be seen in the figure …	**As it can** be seen in the figure …
1	**As mentioned** above…	**As it is** mentioned above …
2	This is not true, **as it is / because it is / since it is** impossible to prove that X = Y.	

32.7 both ... and, either ... or

These expressions are frequently confused, thus leading to ambiguity for the reader:

1. *both .. and* is inclusive
2. *either ... or* is exclusive
3. *both* is only used with *not* in order to highlight a contrast
4. *not ... either ... or* = none of them

	Yes	No
1	We can go to **both** New York **and** Los Angeles. = *2 places*	We can go to **either** NY **and** LA.
2	We can go to **either** NY **or** LA = *1 place*	We can go to **either** NY **either** LA.
3	We **can't** go to **both** NY and LA, but only to NY. = *1 place*	
4	We **can't** go **either** to NY **or** LA. = *0 places*	

Notes:

The position of the preposition changes the meaning:

> We had fun **in both** the parks we visited and also the museums. [two parks and an undefined number of museums]
>
> We had fun **both in** the parks and the museums. [an undefined number of parks and museums]

32.8 *even though, even if*

even though – for real situations. It can generally be replaced with *although*

even if – for hypothetical situations and is generally followed by the simple past or past perfect

Yes	No
Even though / Although the system is designed to work on PCs, it also works on Apple computers.	**Even if** the system is designed to work on PCs, it also works on Apple computers.
Even if we **had** all the time in the world, we would never be able to finish the project.	**Also if** we had all the time in the world, we would never be able to finish the project.

32.9 *in case, if*

Note that:

1. in user manuals, *in case* is generally followed by *of*. *In case of* means *in the event of* and is generally used to refer to emergency situations
2. otherwise, using *if* is generally correct. It means *in cases where*

	Yes	No
1	**In case of** failover / shutdown / blackout, restart your computer.	
2	**If** the value is less than zero, this means that ..	**In case** the value is less than zero, this means that … **In the case that / In the case in which** the value is less than zero, this means that ..

32.10 *instead, on the other hand, whereas, on the contrary*

You cannot use *instead, on the other hand, whereas,* and *on the contrary* indiscriminately:

1. use *instead* at the beginning of a sentence to resolve a problem stated in the previous sentence
2. you can replace on *the other hand* with *whereas*, but *whereas* cannot be used at the beginning of a sentence in a manual
3. do not use *on the contrary* in a technical document. It is only used to contradict what another person has stated

	Yes	No
1	Do not join two independent clauses with a semicolon. **Instead**, make two simple separate sentences.	Do not join two independent clauses with a semicolon. **On the contrary**, make two simple separate sentences.
2	Italian and Spanish are similar languages, **whereas / on the other hand** German is completely different.	Italian and Spanish are similar languages. **Whereas** German is completely different.
3	Italian and Spanish are similar languages, in fact they both derive from Latin. German, **on the other hand**, is derived from ..	Italian and Spanish are similar languages, in fact they both derive from Latin. German, **instead / on the contrary**, is derived from …

THE AUTHOR

Adrian Wallwork

I am the author of over 30 books aimed at helping non-native English speakers to communicate more effectively in English. I have published 13 books with Springer Science and Business Media (the publisher of this book), three Business English coursebooks with Oxford University Press, and also other books for Cambridge University Press, Scholastic, and the BBC.

I teach Business English at several IT companies in Pisa (Italy). I also teach PhD students from around the world how to write and present their work in English. My company, English for Academics, also offers an editing service.

Contacts and Editing Service

Contact me at: adrian.wallwork@gmail.com

Link up with me at:

www.linkedin.com/pub/dir/Adrian/Wallwork

Learn more about my services at:

e4ac.com

ACKNOWLEDGEMENTS

I would like to thank Serena Borgioli, Melanie Guyot, Richard McGowan, Anna Southern and James Wynne.

Index

This index is by section number, not by page number. Numbers in bold refer to whole chapters. Numbers not in bold refer to sections within a chapter.

A
Abbreviations, **18**
Above mentioned / as mentioned above, 23.4
Acronyms, 17.6, **18**
Active form, **30**
Adjectives
  position of, 10.6
Adverbs
  position of, 10.9–10.14
Allow, 32.1
Also, both, 29.5
Ambiguity, **12**
And, 32.5
Apostrophes ('), 16.1
As, as it, 32.6

B
Both … and, 32.7
Bullets, **19**
By vs. *thus* + gerund, 28.4

C
Capitalization, 15.2, **17**
Captions to figures, 20.4
Colons (:), 16.2
Commas (,), 16.3
Comparative form, **25**
Contact details, 7.3
Countable and uncountable nouns, *see* Nouns

D
Dates, **21**
Direct objects and indirect objects
  position of, 10.4
Dots (…), 22.4

E
E.g., i.e., 22.2
Either … or, 32.7
Enable, 32.1
Etc, 22.3
Even if, 32.8
Even though, 32.8

F
Figure, table, appendix etc, 10.7, 17.8, **20**, 23.2, 30.2
For example, 22.1
Forward slash (/), 16.8
Function, functionality, feature, 32.2

G
Genitive, **27**
Gerund, **28**
Getting started, 3.2
Glossaries, 1.7
Google Translate, **13**
Grammatical consistency, 14.6, 19.8

H
He, she, they, see Pronouns
Headings, **15**
Hereafter, 23.5
Hyphens (-), 16.4

I
If, 32.9
If and *when* clauses, 12.3
In case, 32.9
Infinitive, **28**
-ing form, **28**
Installation, 3.1
Instead, 32.10
Instructions, **4**

Introduction, 1.4
It, this, 31.5

K
Key features, 2.1

L
Latin, 12.4
Layout, **14**
Let, 32.1
Login vs *log in*, *start-up* vs *start up*, *etc*, 32.4
Long sentences
  avoiding of, **9**

M
May might, *can* and *will*, 12.7

N
Negations, 10.15, **29**
No one vs. *anyone*, 29.3
Noun + noun sequences, 10.5
Nouns
  countable vs. uncountable, 26.3, 26.4
Numbers, **21**

O
On the contrary, 32.10
On the other hand, 32.10
One, *Ones*, 31.6

P
Parentheses ( ), 16.5
Passive form, **30**
Past participles
  position of, 10.8
Percentages, 21.6
Periods (.), 16.6, 21.4
Permit, 32.1
Procedures, **4**
Product overview, 1.4
Pronouns, **31**
Punctuation, **16**, 18.4, 21.4

R
Reader perspective, **8**
Recommendations, **6**
Redundancy
  avoiding of, **9**

S
Semicolons (;), 16.7
Specifications, 1.6
Spelling, **24**
Subject
  position of, 10.2
Superlative form, **25**

T
Table of contents, 1.2
Tables, 20.5
Terminology, **11**
That, 31.7, 31.8
The following, 23.3
The former, *the latter*, 12.2
(the) last, *(the) next*, 32.3
Title, 1.1
Translation
  automatic, **13**, *see also* Google Translate
Troubleshooting, **5**

U
Uncountable and countable nouns, *see* Nouns
Updates, 7.1
Users, 31.4

V
Verbs
  position of, 10.3

W
Warnings, **6**
Warranty, 7.2
We, *us*, *our*, *see* Pronouns
Whereas, 32.10
Which, 12.6
Which, 31.7, 31.8
Who, 31.7
Word order, **10**

Y
You, *see* Pronouns

Printed in Great Britain
by Amazon